Concrete at Home

MUNI

INNOVATIVE FORMS AND FINISHES

FLOORS | WALLS | FIREPLACES | COUNTERTOPS

Concrete at Home

FU-TUNG CHENG

WITH ERIC OLSEN

Photography *by* Matthew Millman

The Taunton Press

This book is dedicated to Mom Cheng and Bernard Maybeck.

The Taunton Press, Inc., 63 South Main Street, PO Box 5506, Newtown, CT 06470-5506
e-mail: tp@taunton.com

Editor: Erica Sanders-Foege
Interior design: Lori Wendin
Cover design: Mari Nakamura
Layout: Lori Wendin
Illustrator: Chuck Lockhart
Photographer: Matthew Millman, except where noted

Library of Congress Cataloging-in-Publication Data
Cheng, Fu-Tung.
Concrete at home : innovative forms and finishes / Fu-Tung Cheng with Eric Olsen.
p. cm.
Includes index.
ISBN 1-56158-682-X
1. Building fittings. 2. Dwellings--Materials. 3. Concrete products. I. Olsen, Eric. II. Title.
TH6010.C534 2005
693'.5--dc22
2004020332

Printed in the United States of America
10 9 8 7 6 5 4 3 2

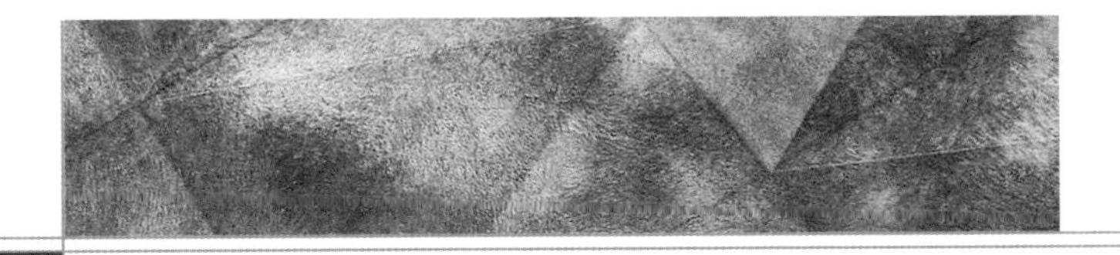

Acknowledgements

To our photographer, Matt Millman, whose inspired images infused my work with transforming light. While Eric Olsen, co-author in the active tense, made sense from nonsense.

To my Taunton editors: Tim Snyder, steady at the helm, and Erica Sanders-Foege, who took charge and got three oars to row as one...also Kathleen Williams and Steve Culpepper.

Special kudos to Hans Rau, a renaissance craftsman, and generous soul, always ready to take on the next challenge with gusto, and to Robert Andrews, the artisan of Cheng Design. Kim Marr, for being there and accounting for my presence, Frank Lee and Margaret Burnett, the always supportive, veteran Chengers, Eric Joost, for steady construction management, Annalyn Chargualaf for her perspective, Howard Hertz, for his counsel. Ching Wei-Jiang, for giving his selfless hand whenever. Winnie W. Yu, partner and friend at Téance/Celadon,

To Dave Condon, fellow concrete artist; to Robert Lawson, engineer, who taught me about rebar and hot sauce; Billie Thomas, contractor and hot sauce king of Culebra, Puerto Rico; and Professor Mark West, for the extraordinary time on that tropical site.

To my clients and friends, and friends who became clients: Wendy Roess, C.B. and Dick Watts, Harry Dent, B. Schuler, Margo Norman and George Triest, Jim and Danielle Sakamoto, Dave and Debbie Trotter, Michael Hogan and Kathy Mayo, John and Polly Marion, Terry McMillan, Ann Hatch, Judith Thompson and Cynthia Brooks, Mitch and Brigitte Durell, Jay and Cheryl Bretton, Robert and Yukari Vincent, Sandra Slater, the Tsuchiyamas, the Ansels, and the Mutchlers. To Mike Miller and his fellow Concretists, Jim Lundy, for the grinding and acid-washing, and Steve Fadelli and his crew for the Clark Place pour.

Thanks to Mari Nakamura, graphic artist; Roger and Frances Williams, of Snow Lion Press, for insights and meals, respectively; to Richard Barnes, for his artistry with the camera, and Robert Ryan and Debbi Beacham for their photo contributions.

Special thanks to Lila Luk, my wife, who really is right, I admit, most all of the time; and to An-Ya, my daughter, who keeps my inner smile intact.

To my late father, Theodore, and to all of my older brothers Alex, Kai B., Carl, Fu-Ding, and the original, vibrant, concrete artist, Ma-Ma Cheng, after four boys—thanks for leaving #5 with enough brains to make a book. Love and bless you all.

Fu-Tung Cheng

Contents

Walls

Fireplaces, Columns, & Architectural Pieces

Preface

When *Concrete Countertops* was published in 2002, we were hopeful that this new approach to kitchen and bath design would find an enthusiastic audience. The book's success really took us by surprise. Tens of thousands of copies are currently in print and interest in concrete countertops continues to grow. The flow of letters, email, phone calls, and trade show contacts has put us in touch with an impressive blend of architects, designers, contractors, and homeowners interested in sharing insights and in learning more about custom concrete.

We responded to this interest by offering concrete countertop workshops for homeowners, do-it-yourselfers, designers, and concrete professionals. Again, the response was overwhelming, which has been of great inspiration to us.

What has become clear is that the impulse to build personal expression in our homes is as inborn and natural as the impulse to grow vegetables we eat in our own garden. The concrete countertop you create can be as unique as you are.

At Cheng Design we have been making concrete countertops for nearly 20 years. But they tell only part of the story. Complete houses, fireplaces, hearths, patios, kitchens, walls, columns, benches —these are the other players. So with *Concrete at Home,* I am honored to show the most exciting projects we have designed and built over the years that go beyond countertops. This book is not meant to be a comprehensive technical manual, but I do hope that we consider the role of good *design* before casting, pouring, or staining willy-nilly. The ultimate success of any project is how well it fits in composition, form, color, and proportion to the space that it resides. I hope the pages ahead will give you ideas that open doors, expand vision, answer questions, and, perhaps, inspire new projects and even businesses—all driven by your own creativity.

–Fu-Tung Cheng

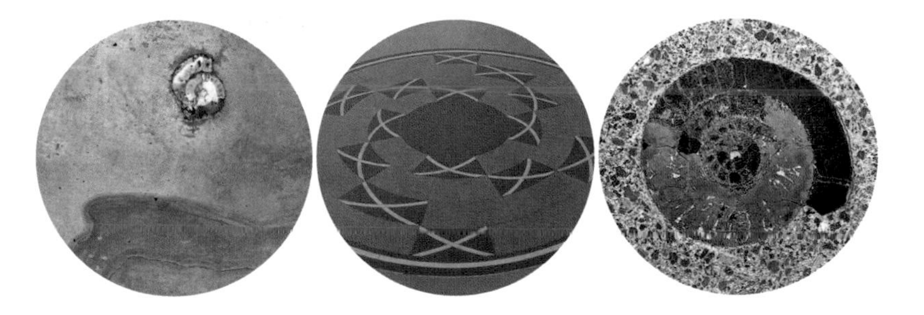

INTRODUCTION

THE AHWAHNEE HOTEL was conceived and constructed in concrete to protect it from wildfires in Yosemite Park.

INSIDE, THE AHWAHNEE LOBBY is a masterpiece of concrete design and craftsmanship. Acid-washed floors and patterns of inlaid cut-linoleum have inspired (and withstood) millions of visitors for over 75 years.

The Ahwahnee Hotel sits in a meadow at the northeastern end of Yosemite Valley. Built in 1926, it is one of the crown jewels of the National Park system. Nearby, El Capitan, a 4,000 ft. granite carapace, rises straight off of the valley floor while Yosemite Falls, a white ribbon in slow motion, cascades down. In the distance, high over Half Dome, clouds return home. I first saw the Ahwahnee one late spring day in 1981. I had wandered onto a trail through a stand of Douglas firs, and came upon the entry to the hotel. Under a redwood timber canopy, cars pulled up, valets opened doors, and women and men in fancy clothes stepped out. Curious, I followed them into the lobby.

When I entered—first thing—the floors caught my eye. They were buffed and waxed to a worn-saddle finish. Broad expanses of veined color were cut-in with beautiful incised patterns, as though etched with a tool. Stone? I wondered. I bent down for a closer look, but couldn't figure out how they were made, or of what. Finally I was amazed to realize, *these floors are concrete*.

When I stood to look around, I saw concrete everywhere, used as I had never seen it used before. I had to know more, so I signed on for the guided tour. We were told that the structure, to reduce the vulnerability to fire, contained little wood. Concrete had been cleverly adapted as finish material to the floors, fireplaces, walls, and beams.

THE PAINTED DETAIL WORK OF Native American themes are seen on concrete beams, columns, and floors throughout the Ahwahnee.

A PENTHOUSE DESIGNED by the author reveals a contemporary connection to the work of Maybeck, Schindler, and Wright. Concrete floors, walls, and countertops form the enclave of the kitchen.

But what really impressed me was how the concrete work merged with the artwork. In the foyer, bold geometric designs reminiscent of Native American baskets and Santa Clara burnished pottery, were inlaid into rust-colored acid-stained floors. In bands on the immense concrete beams, in the drapery, in the ironwork—native graphic designs resonated with the architecture. (Our guide, with some irony, reminded us that the native people who once roamed the valley and inspired these designs no longer made their home here.)

Each time I visit the Ahwahnee, I am refreshed. Strangers become community there, share the pleasures of great meals in a setting of great design, and enjoy the beauty of the public space and the common decency of good government. I am inspired by nature and the nature of human creativity. I head out, eager to pursue the elusive balance among art, architecture, and concrete at home.

A History of Innovation

Although concrete in some parts of the Ahwahnee was made to look like wood and stone, concrete's history has more to do with innovation than imitation. Architects, designers and builders have been experimenting with concrete's structural and

BERNARD MAYBECK'S WALLEN HOUSE in Berkeley, Calif. His innovative concrete application influenced Julia Morgan (Hearst Castle) and a generation of designers since.

sculptural qualities for more than a century. Today, you'll find countless examples of their work in any major city—cast columns, balustrades, and Art Deco façades and tableaus.

In California, where I live, we're fortunate to have concrete homes designed by Frank Lloyd Wright, Rudolf Schindler, and Bernard Maybeck—three architects known for their innovative use of the material. Rather than mask concrete's natural tone and texture, these pioneers preferred to let the material speak for itself, retaining the marks of form boards, the modularity of cast blocks, and the obvious mass of solid walls. Having this heritage of well designed work in the neighborhood definitely made an impression on me that influenced the outcome of many projects throughout this book.

When I first traveled to Europe, I was surprised at how much I loved all the stuff on buildings that I was taught to detest in art school. Expecting much form and little function, I wasn't at all prepared for the beauty I encountered in Bernini's fountains or Gaudi's undulating concrete walls. I realized that centuries of culture could and should be crafted into buildings. I saw the ingenuity of masons and sculptors passing down a tradition of design through the building process. Art theories and learned concepts gain traction with hands-on experience: I discovered the design possibilities

THE KITCHEN of this San Francisco loft (Natoma Street) illustrates that craftsmanship is just one essential to producing good concrete design.

of concrete by mixing some up and playing with it. This is the best way I have of explaining the connection between what I saw at the Ahwahnee and in Europe with what I do today.

Understanding the Past to Create the Future

I've often wondered why so many look to the past for design inspiration. What vital ingredients are missing from today's houses? I think it's the soulful, the simple, the elegant; a quality of craftsmanship, design, and respect for materials that we associate with work from other eras. Today, we associate these qualities with the styles of the past, but rather than simply mimic the style of the past, we have an opportunity to invigorate our contemporary homes with expression and creativity on contemporary terms. Christopher Alexander's *A Timeless Way of Building* eloquently encourages us to rediscover the "pattern language of designing and building instinctively" and not allow ourselves to drift ". . . in superficial trends and style, without a cultural guiding hand."

THE LIBERAL USE of concrete in many guises continues in the loft's living room floors and hearth. Hand troweled finished floors are separated by copper strips, with a diamond ground border inlaid with fossil punctuation designs.

Thinking of style as an assortment of possibilities rather than as a prescription or recipe from the past is actually quite freeing, and it's suddenly much easier to understand how concrete might fit in just about anywhere.

In our own work, we try to take the time to find new ways to use this age-old material in the context of the architecture and find the way toward soulful, personalized, well crafted design that evokes the same feelings of comfort and value found in homes built hundreds of years ago.

CONCRETE AT ITS BEST. Berkeley's Edwards Stadium, built in 1932, features work in concrete that epitomizes the Art Deco style.

THE FUNDAMENTALS
1 of Floor Design

PART ONE

A good floor is a reflection of patterns of circulation and rest in space. As we move through any building, there is a rhythm and pace that is influenced by the nature of the floors and their scale. Each material and the way it's composed can dictate the feeling of the room. Antique wide pine planking, with its natural gaps, for example, reads "rustic, homey, country." An intricately inlaid marquetry oak floor, on the other hand, conveys the aesthetic of the Craftsman-style house. Softwoods like pine, fir, and poplar reflect their temperate origins; hardwoods like teak, mahogany, and bubinga conjure up the tropics.

The beauty of concrete as a flooring material is in its receptivity to any design treatment. It can accept a variety of textures and inlays. It can be carved, ground and polished, sand-blasted, stamped, stained, drawn on, painted, or dyed. It can evoke any feeling or allude to any time or place or style, from contemporary to ancient, from slick and commercial to warm and rustic. No other flooring material presents as many opportunities for creativity—or, arguably, creative excess—and that's the topic here.

Most of the design treatments we'll discuss provide opportunities for the artist-designer, architect, artisan, or

« LINE AND FORM DRAW VISITORS to a comfortable resting place by the entrance inside the author's home. A platform of recycled gym flooring steps down to the smooth, contrasting concrete slab.

»THIS BEAUTIFULLY PRECISE floor detail is comprised of linoleum inlaid in concrete at the Ahwahnee Hotel, Yosemite, Calif.

ANOTHER BORDER DETAIL in the lobby floor of the Ahwahnee Hotel reveals how vibrant the color remains after more than 75 years.

homeowner to have a big impact on the final look of a floor, with little added expense and a minimum of specialized skills or equipment. But because whatever you do to concrete will end up literally cast in stone, it's not a bad idea to spend some time with a pencil and some graph paper to work out a graphically pleasing design before you pour.

But what is considered graphically pleasing? With floors it means starting with the basics of good composition: the balance of line, line weight, and proportion with all the various materials and techniques at our disposal. It means using joint lines and inlays with color fields, and stamping patterns in harmonious balance. Repetitive compositions are the easiest to accomplish once a basic pattern is established. Asymmetrical compositions are more intuitive and perhaps more difficult for the novice.

BY DESIGN

A FLOOR HAS TO BE LEVEL, flat and smooth (and contain some minimal texturing to prevent slips if it's outdoors). But there's a lot you can do design-wise within these constraints. A floor is a two-dimensional composition, so for inspiration, consider the designs of people who work in two dimensions—those who do fine wood floors, rug-makers, or graphic artists.

Safe Steps

Ideally, when designing a transition from one level to the next, design two or more steps. A single step is more difficult to see than two or more. If one is the limit, make sure the adjacent surfaces are different textures or colors so the transition is obvious. Building codes in most areas specify 4 in. as the minimum height for a step from one level to another. Anything less is considered a tripping hazard. Consider gradually ramping the floor from one level to the next as an alternative.

THE STEPS DEFINE this farmhouse entryway, giving importance to the space. From here, by stepping up, you may enter the living room, bedrooms or even the back yard.

Changing Levels

A change in the level of a floor can signal that you're moving into a new space that has a different purpose or function, and where a new set of behaviors may be expected. In a traditional Japanese home, for instance, in which the division between inside and outside is typically blurred, a simple elevation change at the transition marks the point beyond which shoes aren't worn. Such level changes are especially effective when there are no other markers, such as a wall or screen or doorway, to separate one space from another. Level changes can create an open space that is scaled for comfort and a variety of purposes.

For best effect, the level of the floor should work in tension with the ceiling above it. Frank Lloyd Wright talked about compression and release as essential design elements: You might compress the space in an entryway, for instance, by lowering the ceiling and raising the floor, and then opening up the space beyond the entry with a step down into the living room under a high ceiling. The interplay of ceiling and floor heights helps define the experience one has in the space.

NATOMA STREET

The Natoma Street project in San Francisco, a large penthouse interior we designed and built, has a single-level concrete subfloor and a high, arched ceiling, which gave us plenty of room in which to play with elevation changes in both the floor and the ceiling. We raised the floor of the entire central area of the penthouse with a subfloor of plywood, sound-board, and a 2½-in. concrete slab (see Chapter 2 for more on this). (A hydronic heating system was installed directly into the slab.) The floors for the entryway, kitchen, and bedrooms are on a lower level. The level changes define spaces and add interest. In some parts of the penthouse, the transition is a single step; in others it's a ramp.

THE CENTRAL AREA of the penthouse floor is a raised concrete platform, which acts as a transition from public to private space—here we look from the master bedroom to the living room. In the far corner of the room, a special platform delineates space for yoga.

USING A LEVEL CHANGE as an accent, the platform in this contemporary farmhouse complements the adjacent concrete slab and supports a wood-burning stove.

IN THIS CONTEMPORARY California home, the concrete curb acts as a toe-kick to the cabinetry and appears as a natural extension of the floor.

Lesson in Composition

For this Sebastopol, Calif., home, reminiscent of a Japanese farmhouse, it was essential to highlight the experience of entering the home. Instead of designing a seamless transition from outside to inside, I created an entry that is full of level changes and detail that give visitors the feeling of interconnectedness with the deep structure of the house. Because the living area is an open floor plan, this entry space is an important transition. It's a place where visitors can pause and take in their surroundings. The careful balance of shape, detail, and materials creates a space in which function and flow are completely integrated.

» **IN THE FOYER,** the acid-washed concrete step is keyed into the adjacent walls. Inlaid flooring becomes a platform that, flanked by concrete and Douglas fir columns, divides the space into kitchen (left) and living room (right).

GUESTS STEP DOWN to a cozy foyer before entering the expanse of the main living area. A crescent of recycled granite ornaments the neutral concrete floor.

^ **IN THIS MORAGA, CALIF., HOME,** the intense gray-green floor works to deepen the richly-hued stucco walls. To achieve the final color, the floor received integral black pigment and an acid wash stain. The patch of wood draws the eye to the floor's intricate pattern.

« **THE INSPIRATION FOR** the entryway above came from the Ahwahnee Hotel dining room, with its subtle variations of grays and greens and sharply delineated fields of color.

Color

Designers and fabricators have at their disposal a variety of concrete coloring agents: the old standbys such as integral and broadcast pigments and acid washes, plus a host of new epoxy and acrylic dyes, stains, and paints (see Chapter 3). These days there are few limits on the colors that a concrete floor can sport.

Often you'll find flamboyantly colored concrete floors in restaurants, shopping malls, and casinos. In a commercial setting, floors are meant to excite the senses. In the home, you may want a "calmer" floor, one that tends toward a natural palette. And because the floor color is often one of the first that is established in a room, it makes sense to go for neutral tones so you have more choices elsewhere.

Our concrete floors are almost always made with natural gray concrete (no white cement, no brilliant epoxy paints), natural aggregates, and pigments that yield natural hues. We shy away from bright or clearly artificial colors, which may appease a momentary trend but lack a "timeless" quality. For added visual interest, we rely on elevation changes, lines, inlays, and textures.

<< **THE ORINDA MOVIE THEATER,** circa 1930, in Orinda, Calif., with its classically expressive terrazzo floors, is one of many Art Deco jewels that set the precedent for the opulent floors seen today in commercial concrete applications.

TERRAZZO

Terrazzo was invented in the 15th century by Venetian stone masons who used the left-over chips from marble carvings to pave their terraces (*terrazzo* is Italian for "terrace"). They embedded the chips in a clay matrix and then ground the

>> **FIELDS OF COLOR** are sharply defined in terrazzo floors, allowing for intricate patternwork.

rough surface with stones, a time-consuming, back-breaking process. To bring out the color of the marble, they sealed it with goat's milk. Wooden strips—later removed and filled with mortar or cubes of marble called mosaic tessarae—were used to divide fields, and patterns and pictures were created.

Most terrazzo floors used in commercial settings today are made with an epoxy matrix rather than concrete. These epoxies create vivid color patterns across large floors. Our ground and polished concrete floors are, technically speaking, terrazzo. That is, they're patterns of stone set in a concrete matrix that has been polished. Commercial terrazzo is uniform in color and aggregate size. By contrast, ground concrete will look more natural, showing the variegation of the aggregates, depending on how close to the surface the material ends up after screeding and troweling.

ACRYLIC AND DYE

In contrast to terrazzo, there are new "painterly" methods for coloring concrete, which produce a completely different spectrum and look. Michael Miller, a concrete design specialist, uses acrylic paint and dye to create hues and textures that allow the concrete to take on vivid colors and effects. Topical sealers protect against wear. (We'll learn more about him and the Concretists' work in Chapter 3).

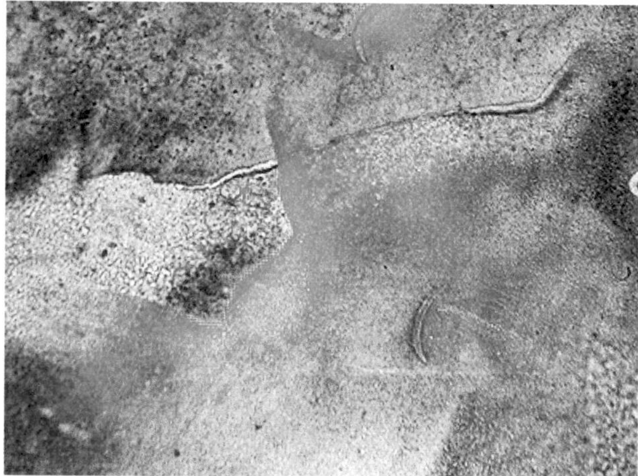

CONCRETISTS MICHAEL MILLER AND DANA BOYER created this intensely colored floor by spotting the cementitious overlay with dry color, followed by stenciling, spray-painting, acid-washing, and applications of dye. The entire floor was then sealed with an acrylic sealer.

THE 75-YEAR-OLD INTEGRALLY-COLORED red iron oxide concrete floor of the Téance Tea Room/Celadon Fine Teas in Albany, Calif., was given a second life after it was stripped of adhesive, then ground and polished.

THE OWNERS OF THIS Occidental, Calif., home planned a continuous floor from their kitchen to the patio, but wanted each square of the entire grid poured individually. Though it took longer to achieve, the effect is one of oversized hand-crafted tiles.

Lines

Most lines in a concrete floor are placed there as crack-control joints. These deep grooves, tooled into wet concrete or cut into cured concrete, create deliberately weakened pathways along which cracks will likely travel. Crack-control joints are typically laid out in a regular pattern of evenly spaced lines or in a grid, which cracks often as not will blithely ignore as they form. We treat control joints as a composition tool that can define space, separate color, or reinforce patterns.

BY THE WAY

Joints that have been tooled into wet concrete will tend to have a rounded edge. The effect is somewhat "rustic." Joints that have been saw-cut into green or cured concrete will have a sharp edge that is more "industrial."

Inside-Outside Concrete

There's something elegant and pleasing about a floor that passes the threshold from inside to outdoors on one level. Concrete lends itself particularly well to this application—the design options are numerous and it stands up well to the elements. For the entryway to this home in Oakland, Calif., I poured a single concrete slab that extends from the outside into the foyer.

After all, bringing nature into the home is at the heart of indoor-outdoor design, so I embedded a stone into one corner of the slab. This, to my thinking, was a way to integrate a sense of organic randomness into a rectilinear slab. Also, I applied a salt finish for texture, then acid-washed the concrete palm green to make the slab appear more natural. For the control joint, too, I created a graceful, organic shape.

THE CURVED CONTROL JOINT is created with a single brass strip and leads the eye right outdoors.

AT THE INTERSECTION of rock and concrete, a profusion of texture. The jagged edge contrasts with the neutral section and with a field of green concrete, which was treated with a salt finish to produce the stippling effect.

THE GREEN SWATH of acid-washed concrete welcomes you inside this California contemporary. In comparison, the broomed, unstained section shows just how much color the acid-washing process can produce.

Inlays, Textures, Broadcast Stone, and Stamping

There are design options—inserting inlays, broadcasting stone or color, or adding texture or stamped patterns—that can be done after the concrete is poured.

INLAYS

Over the years, we've stuck all sorts of things into our concrete floors: cut rocks, marble and granite pavers, flagstones, tiles, metal strips, coins, car parts, and what has become a trademark of Cheng Design, cut and polished fossils. They are a relatively easy way to impact the look of what may otherwise be a featureless floor.

TEXTURES

Probably the most common texture used outside is the brushed or broomed surface, which is created using a special broom that's dragged or pushed across the still-wet concrete to give it some "tooth." You can create checkerboard or herringbone patterns or weave the brush from side to side.

Wash finishing is done when the concrete is still green but hard enough to walk on without leaving a footprint. Some of the top layer of cement, called cream, is washed and brushed away. This leaves the fine sand in the mix exposed and gives the slab a non-slip surface.

A salt finish is produced by scattering and then troweling rock salt or nitrogen fertilizer pellets onto the surface of the wet slab. After the concrete has hardened, the salt or pellets are washed out, leaving a stippled surface. You can also produce this effect, albeit in a repetitive pattern, with a rubber roller.

THE LINES ON THIS FLOOR are not control joints, but decorative, to add texture and graphically divide the slab.

MOST THINK OF WEEDS sprouting out of pavement as a sign of neglect. Here, the notion is turned on its head as the concrete embraces nature. A salt and stained finish, with brooming on one side to add texture, provides a safe surface.

BROADCAST STONE

Decorative aggregates such as pebbles, colored rock, marble, or glass are spread onto the wet concrete. As the concrete takes its first hard set (a few hours after the last troweling), the entire slab is lightly scrubbed with brushes and washed to reveal the decorative material. For interiors, the concrete cures, then can be ground and polished to reveal the aggregates.

STAMPING

Stamping concrete is an affordable way to imitate the look of natural stone or brick without having to pay for the real thing. Proprietary rubber patterns that create effects resembling natural stone, and preconfigured patterns such as brick or cobblestone, are commercially available, but tend to look like what they are—artificial.

My stamp impressions are full of variety and often are made from found objects (see Chapter 3). Used with discretion, a stamp is an effective decorative tool. The concrete can simply be concrete, without trying to imitate something else.

But this is often the case with good design. And though there are no limits to the imaginative combinations of the practical and the aesthetic, applying a philosophy of restraint is a good idea when using any of the techniques we've discussed here to design your own floor.

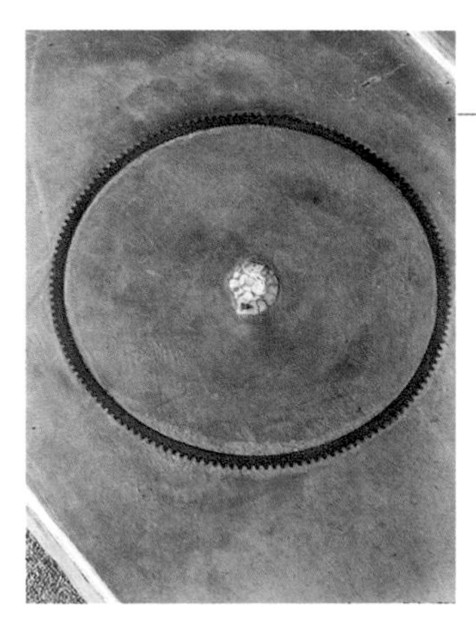

BY THE WAY

Adding inlays is an opportunity for the homeowner to have a big impact on the final look of the floor. Experiment. Take the time to lay out a graphically pleasing design on paper and then mock it up, or part of it, to full size to get a feel for the effect.

FLOOR ESSENTIALS:

2 Preparation, Execution & Pouring

Concrete slabs have traditionally served as underlayments for other materials, and thus we've grown accustomed to having our concrete hidden—and ignored—under carpet, linoleum, wood, tile, or marble. But with a fresh perspective and a few new techniques and simple materials, you can cost-effectively turn what might have been an ordinary concrete slab into something stunning and long lasting, something you'd never want to hide.

There are two basic types of concrete floors: the slab on grade (interior or exterior) and the interior slab over a subfloor. A slab on grade (or a ground-level) concrete floor is the most common. And it's probably a good first project if you're a committed do-it-yourselfer with limited experience in working with concrete. Logistically, this type of floor can be a small, simple project. You don't have to fret about out-of-level subfloors, over-spanned or inadequate floor joists, or excessive or awkward elevation changes from one surface to another, all problems you may encounter when pouring an interior slab over a subfloor. You do need to consider the condition of the underlying soil, however.

« THE FLOOR IN THE ENTRYWAY of this new Rancho Santa Fe, Calif., home complements the graphic quality of the space—open rafters and celadon plaster walls are held in balance by the concrete terrazzo. The arc of the copper control joint echoes the curved ceiling.

>> **AT THE NATOMA STREET POUR** in San Francisco, the prepped floor was an intricate web of rebar, inlays, tubing and screed guides.

If you're constructing your floor outdoors, there are weather variables to contend with, but often one rather common logistical problem—getting the concrete to the site—is much simpler to resolve: Usually the truck can just back up and unload directly into the forms. Designing and fabricating an interior slab over a subfloor requires a little more technique and preparation. But you don't have to worry about the weather, and curing conditions are usually much simpler inside than out.

BY THE WAY

As a good move environmentally, substitute a sack of fly ash for one of cement per yard of concrete. A recycled material, fly ash is a byproduct of power plants. It's like a pozzolan and less polluting than cement, which requires a tremendous amount of energy to make. Concrete with fly ash is as strong as regular concrete and finishes just as well. It also tends to be more workable. Darker than that made with Portland cement, concrete with fly ash is denser, so it stands up well against water migration.

When you're admiring a beautiful concrete floor, you don't see the many structural details that lie beneath the surface. Yet what you can see and enjoy largely depends on the supporting structure, whether it be wood trusses, or joists and plywood, or earth. In planning, you'll need to consider such things as the slab's thickness, the composition and weight of the concrete, reinforcement details, transitions to existing floors, and even heating and inlay elements that need to be placed before the pour. Below is an overview—not a step-by-step guide—of how to pour a slab. But for those with experience, this chapter will enable you to create a technically sound job, and serve as inspiration for spectacular jobs in the future.

After addressing what goes into the mix, we'll focus on two projects: Clark Place, a basic slab-on-grade floor and patio in El Cerrito, Calif., and the Natoma Street penthouse, a sophisticated, high-end interior pour in San Francisco. Together, these floors provide a fairly thorough overview of concrete floor design and fabrication, present

A TYPICAL ORDER FORM FOR STANDARD READY MIX

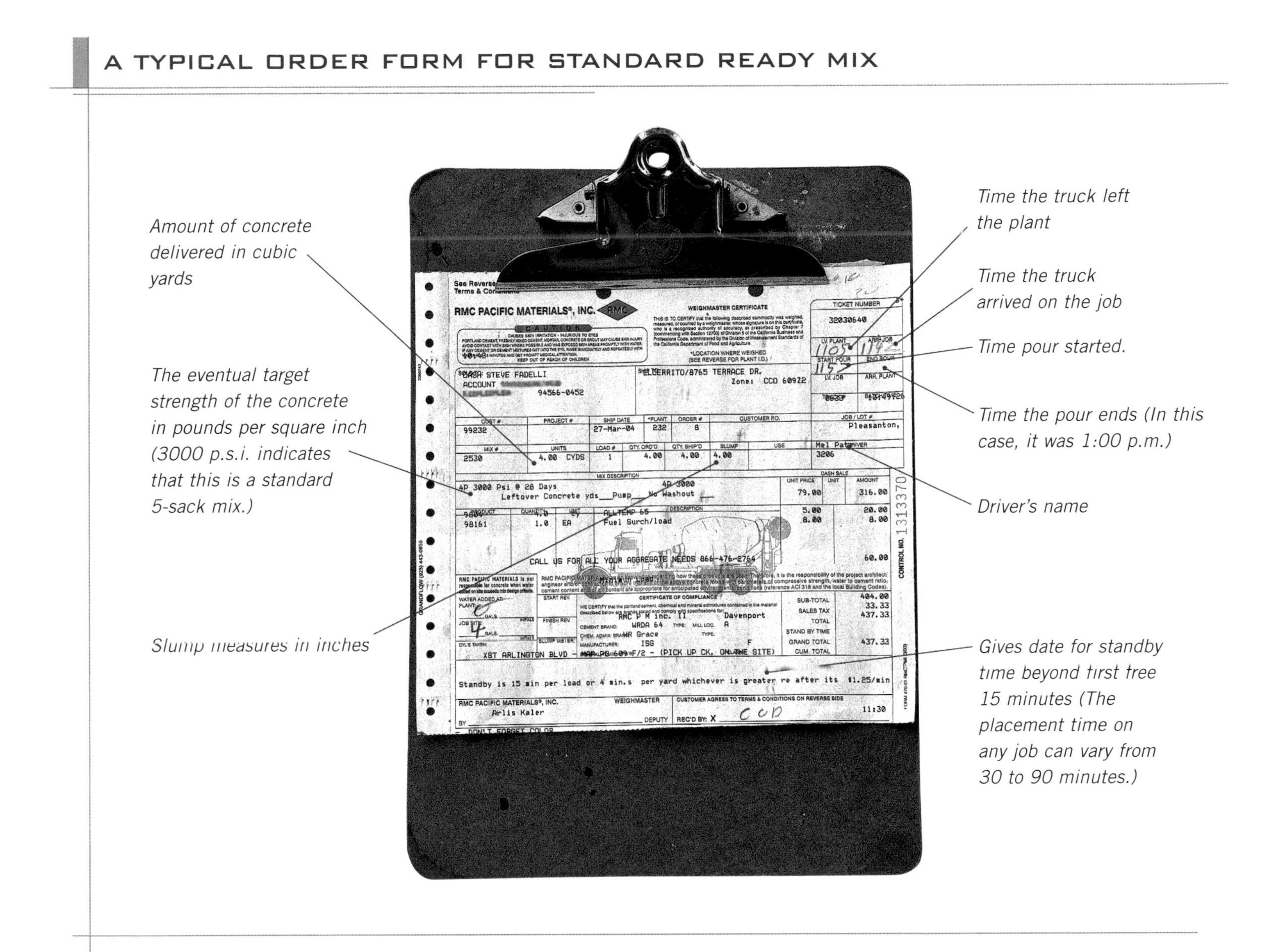

interesting challenges, and illustrate the power of simplicity in good design.

The Mix

Whatever the project, it's never too soon to consider the ingredients for the mix. It's rarely the case that you mix your own concrete for a floor, unless it's a small one. Instead, rely on ready-mix suppliers, which means giving them *very* clear specifications as to ingredients that go into the batch. And make sure there is adequate manpower to screed, trowel, and finish. Once water is added, concrete starts to set up. The mix that gets delivered will affect workability, strength, and final appearance of the finished floor. Make sure to specify the following:

BY THE YARD

Call for 6 to 6.5 sacks (94 lb. cement powder per bag), which is richer than the off-the-shelf 5-sack mix from most ready-mix plants. With more cream—cement and fines—coming to the surface during finishing, the result is a harder, slicker,

The Recipe

To make 1 yard of concrete for walls or floors, mix:

- cement: 6 to 6.5 (94-lb.) sacks, 564 to 610 lb.
- ⅜-in. pea gravel: 1,000 lb. (lightweight expanded shale: 810 lb.)
- fines: 1,944 lb. (lightweight: 1,080 lb.)
- water reducer: specifications vary by manufacturer; we always use the maximum specified.
- polypropylene fibers: 1 lb. (½ the recommended dose)
- pigments: vary according to desired color or shading.
- water: enough with water reducer to create a 6- to 7-in. slump (about ¼ lb. per 1 lb. cement).

more refined finish. Also, it's possible to polish the surface more extensively before getting into coarser aggregates. (Be aware that this mix shrinks slightly more than a 5-sack mix of the same volume.) Also ask for Type II cement, which allows for more time to work the slab—concrete made with Type II cement hardens or "goes off" more slowly than concrete made with Type III or "high early" cement. When pigments are added to concrete made with white cement (also slower to go off), the results can be brighter and more vivid.

AGGREGATES

For floors less than 3-in. thick, specify ⅜-in. pea gravel as the largest aggregate (called, "maximum size aggregate," MSA, or ⅜ in.) For thicker slabs, use the standard ¾-in. crushed rock, which is more difficult to finish than concrete made with pea

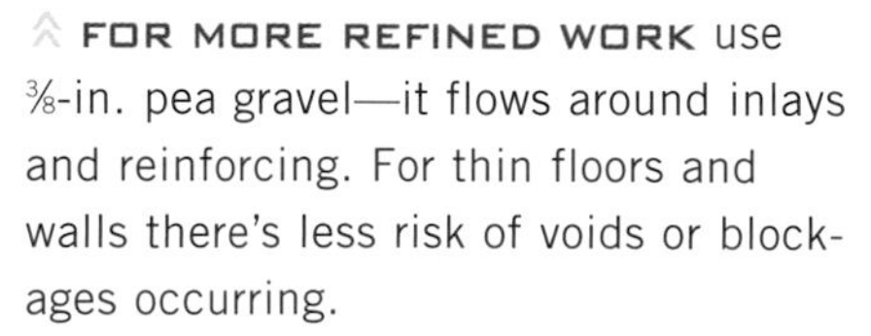

FOR MORE REFINED WORK use ⅜-in. pea gravel—it flows around inlays and reinforcing. For thin floors and walls there's less risk of voids or blockages occurring.

FROM HEAVY-DUTY STAINLESS STEEL to polypropylene for floors, walls and countertops, there are many types of fibers for strengthening concrete.

gravel. Also, it's more difficult to work around reinforcing materials and inlays, but the result is a concrete that is structurally stronger. Lightweight aggregates like volcanic scoria, expanded clay, or shale may be suitable for interior slabs where weight is an issue—they can lighten the load by 30 lbs. per cubic foot.

FIBERS

Fibers strengthen concrete by reducing the risk of "microcracks," and by keeping them from growing when they do form. Use Nycon nylon or Fibermesh Stealth fibers, for example, at half the recommended amount; at full dose, they may affect the look of the finished surface.

AT THE NATOMA STREET PROJECT, the pumper truck had to deliver the concrete four stories up to the penthouse through a 6-in. hose.

PIGMENTS, EITHER POWDERED or in suspension, are used for integral coloring.

BY THE WAY

Call your ready-mix plant to reserve delivery, and then confirm the order the day before. If weather conditions are going to be very hot and dry, you may want to specify a retarder; if they're likely to be cool, you may want an accelerator. If rain is in the forecast, consider rescheduling.

RETARDER OR ACCELERATOR

A retarder slows the rate at which concrete hardens, keeping it workable longer, but when the concrete begins to "go off," it hardens very quickly. An accelerator speeds the rate at which concrete sets up. Those with experience use it in quite cool conditions or when they need little time to finish the slab.

BY THE WAY

There are areas of concrete work where much common sense—but little experience—is all you need. Pouring and finishing a slab isn't one of them. Here experience rules. Hit the slab too early with the trowels and you end up with a weak finish that may crack or spall (flake off). Hit it too late and you end up with a hard, lumpy rockscape. We recommend calling in professional help at least for the finish work.

REDUCER

Water reducer helps lower the risk of cracks. It enables you to use less water but still produces a mix that is plastic and workable.

INTEGRAL COLORING

There are many ways to color a concrete floor, most of which are done after the concrete has been poured (see Chapter 1). If you choose to color the concrete integrally—which means coloring the entire slab—you'll need to specify the amount and color of pigment to be added at the batch plant. You may have to supply the pigments, which are measured as a percentage of the total weight of the cement. Because pigments tend to weaken concrete, their total weight should not be more than 10 to15 percent of the weight of the concrete. Integral pigments can be costly, adding as much as a third to the bottom line. Carbon black is fairly low-cost, though, and we often use some—

» **THE CLARK PLACE PROJECT** in El Cerrito, Calif., represents a manageable pour in a typical situation, one that any owner or builder can tackle.

TOPSOIL WAS EXCAVATED down to compacted clay and an 18-in. deep trench—with 3-ft. piers at the outer corners— was dug around the perimeter of the slab.

THE SLAB WAS FORMED WITH typical 2-in. x 12-in. boards. We screwed the boards to metal form stakes, drove in the stakes along layout lines, then used a laser level to guide adjustments.

FOR A VAPOR BARRIER, a layer of 6-mil plastic sheeting went down. That was topped with gravel and a second sheet of plastic.

½ lb. of carbon black per cubic yard—to darken the concrete, since unpigmented concrete cures to shades of unattractive pale gray.

SLUMP ON DELIVERY

Slump, a measure of wetness or workability, refers to how much fresh concrete, poured into a cone-shaped mold, settles after the cone is removed. On a scale of 1 in. to 12 in., concrete that slumps only 1 in. is very dry and stiff and will be difficult to work; a 12-in. slump indicates extremely soupy concrete that will likely shrink and crack. For floors and walls, we specify a slump (with water reducer) of 6 in. to 7 in. (the consistency of very runny oatmeal). This mix won't stiffen too quickly, leaving time for screeding and troweling. It will also tend to settle evenly.

« **PERFORM A SLUMP TEST** immediately after the concrete begins to come down the shoot to get a precise slump reading and to have a record of the batch. Here, the author confirms the slump and is ready to go.

WE DRILLED HOLES IN THE OLD FOUNDATION, 16 in. on center, and used a 2-part epoxy to glue in lengths of #4 rebar, to which we tied a grid of #4 rebar on 16-in. centers, plus a rebar cage in each pier and along the perimeter.

MAKE SURE THE REBAR is no closer than 1 in. to the surface, because it can cause "ghosting" (discoloration) if placed too close to the surface.

AN EXPERIENCED HAND can estimate the slump just by feel.

Slab on Grade

For designing a slab floor on grade, the condition of the soil and proper soil preparation are essential considerations. A soils report from a civil engineer will tell you what type of foundation for the slab will be best for the conditions. The report will also outline appropriate insulation and drainage.

The slab for the Clark Place job was designed for a small, 10-ft. x 12-ft. addition to the ground-level basement of the co-author's house, which was built nearly 80 years before on a trench foundation laid down on expansive clay soil in the East Bay hills outside of San Francisco, a very active seismic zone. Thus, we needed a slab designed to minimize cracking due to soil movement that would also move as a unit with the rest of the house. We opted for a "raft" design, a 6-in. slab tied to a perimeter foundation, a simple slab that will be quite adequate for the conditions.

3 **A LAYER OF 1-IN. RIGID FOAM** insulation was installed on the second sheet for a hydronic heating system.

4 **THE HYDRONIC TUBING** was tied to the rebar so it's centered in the slab. The tubing connects to a manifold, which will be tied in to the home's water heater.

PLACING AND FINISHING THE CONCRETE

Following are the basic steps in placing and finishing concrete, up to the point that it's time to manipulate the surface: placing inlays, tooling or cutting in control joints, and broadcasting color or rocks. The following is general information so that you can monitor the progress of a contractor's crew, or so that you know the fundamentals involved if you plan to be the contractor and direct the professional finishers yourself. These steps also apply to interior slabs.

Remember that once water has been added to the dry concrete mix, the clock starts to tick. The weather is a factor, too. On a hot, dry day, you'll have less time to work the concrete, so you may need a bigger crew than if the weather's cool and damp.

POSITIONING THE REBAR is crucial to the structural integrity of the finished floor.

For a typical slab on grade, placing the concrete will likely be a simple matter of backing the truck up to the form and running it down the chute. Hard-to-reach sites may require a pumper and hose. In either case, if you've placed inlays or internal support like rebar or remesh in the form, be careful that you don't dislodge the inlays while placing the concrete.

Vibrating. Vibrating drives out air bubbles and "liquefies" concrete so it flows completely around the rebar and any inlays, preventing voids that can weaken the concrete or cause unsightly holes where exposed. We used a "stinger" or stick vibrator on the Clark Place job. Some contractors also use a tamper—also called a "jitterbug" (especially if ¾-in. crushed rock is used in the mix), typically a large, flat piece of perforated metal with handles.

ON THE CLARK PLACE PROJECT, we used ¾-in. crushed rock in a fairly stiff mix, so thorough vibrating was essential. If fine detail were in the plan, we'd opt for ⅜-in. pea gravel.

A SLAB THIS SMALL can be screeded with a straight 2x4 by sawing it across the surface.

Screeding. Sufficient and careful screeding will have a major impact on the flatness of the slab by leveling the concrete and filling dips. Screeding is done by pulling the screed—a straight 2x4 or an aluminum box beam—along the top of the form or along guides while moving it side to side in a sawing motion. On smaller floors, screeding begins once the entire floor has been poured. For a larger pour, it begins while the concrete is still being poured. Insufficient or sloppy screeding can leave bulges or dips that, even if too subtle to see when the concrete is still wet and being worked, can nonetheless be a problem later, especially if you're trying to grind the floor for flatness.

WE PLACED FOUNDATION BOLTS around the perimeter of the slab at 12-in.intervals. The bolts were hooked at the bottom, then wrapped around and wired to the rebar.

EACH FOUNDATION BOLT was wriggled in place to make sure it was properly seated.

A BASIC SLAB ON GRADE

A monolithic slab is the easiest and most economical type of slab to pour in areas where winters are mild (and the thickened edge of the slab need not extend deep into the ground). Perimeter and under-slab insulation should be added if living space will be built atop the slab.

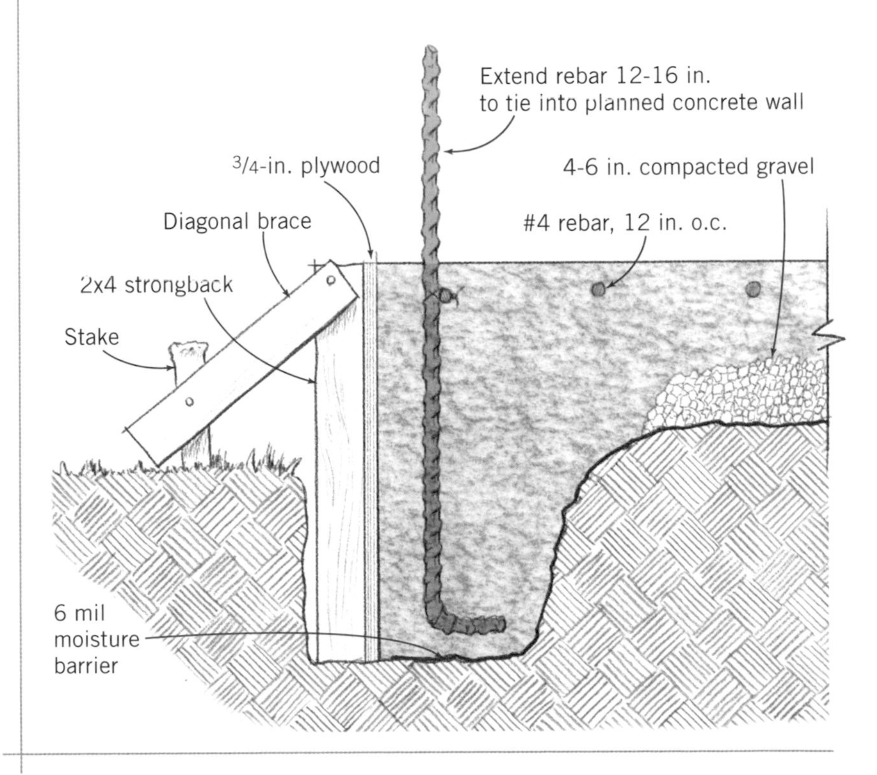

THE BULL FLOAT SHOULD SMOOTH the surface without digging into it. The process can reveal any areas that need more concrete. It's crucial to avoid dragging concrete from one area to another.

Be careful not to bury any inlays. If the inlays have been set properly, and if the screeding brings the surface of the concrete to the level of the forms, then the inlays should be covered only by a very thin film of cream.

Floating and troweling. Fight the temptation to hit the concrete too early when floating and troweling, and don't work the concrete excessively. As a rule of thumb, if you find that you're pulling lots of water to the surface at any stage, back off. Excess water (a puddle, not just a sheen) raises the water-to-cement ratio in the thin top layer of cream. Too much water will weaken it. Over time, this compromised surface of fines and cement may begin to show excessive wear and may even spall, or peel off in thin flakes.

After screeding, the first pass over the concrete is done with a bull float, a wide blade of metal on a long handle. The bull float pulls water and the cream to the surface, filling the small voids, and helping to smooth the surface, which, when the floating is complete, will be covered with a thin

WHEN TROWELING, hold the trowel as flat as possible, otherwise, you risk creating a "washboard" surface. On later passes, as the concrete hardens, you can angle the trowel slightly.

THE FIRST PASS with a jointing tool pushes aggregates out of the way. As the concrete is troweled, the joints may fill with cream, so they'll need subsequent passes with the tool.

BY THE WAY

In the industry, a concrete floor is considered "flat" if there is no more than a ⅛-in. variation across 10 feet. Waviness can, to a point, be ground out, but heavy grinding can change the color or the look of a finished slab by uncovering the larger aggregates. And you may reveal ghosting of reinforcement materials in the process of trying to even out your surface.

sheen of bleed water. The concrete is not yet quite strong enough to support a person on kneepads, and is still too soft to receive inlays.

Bull-floating is followed by a wooden float or darby, after which the concrete should be firm enough to receive tooled-in control joints, inlays, a color hardener, broadcast stone, or stamped impressions (turn to Chapter 3), if any of these are in the plan. If not, then work the surface with a magnesium and/or steel trowel. Troweling continues until the concrete is very hard, usually for several hours, depending on how smooth and "polished" a surface you want (the more thorough the troweling, the less grinding and polishing later, if that's in the plan).

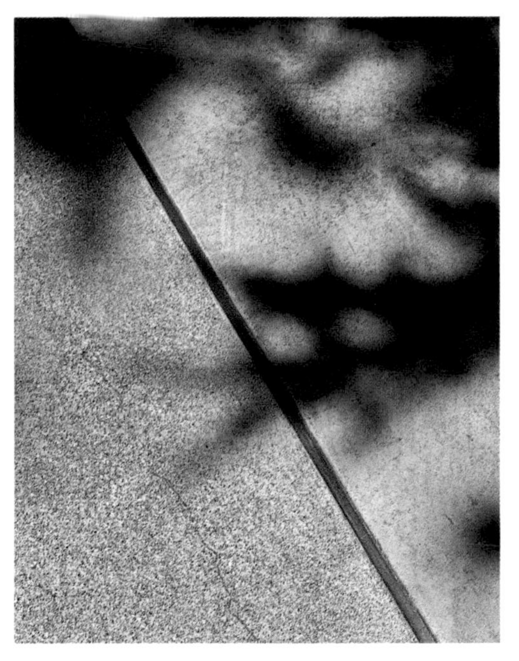

COPPER INLAID STRIPS were used for the control joints on this project.

HERE, ONE AREA OF THE SLAB was in shade and another in sunlight, so a tarp was needed to ensure the concrete cured evenly.

BY THE WAY

It's possible to keep the elevation change between floors to a minimum by replacing the subfloor in the room where the concrete floor will be poured. For example, if your 2-in. concrete floor butts up to a ¾-in. wood floor, the difference in height can be a stumbler. Attach the new subfloor between the joists (use cleats). This will lower the new floor to the thickness of the old subfloor, typically ¾ in. A 1-in. concrete floor will be just ¼ in. higher than the adjacent wood floor, a slight differential that can be masked with a trim strip.

Control joints. Cracks are more likely to form in a large expanse of undifferentiated concrete than in a small slab or a large slab divided into small sections by control joints. Control joints, again, are grooves that are cut or tooled into the concrete. A control joint is like a line scored in a tile; it creates a weak area on the surface, which any cracks that form may follow. A control joint needs to be at least a quarter of the thickness of the slab and is most effective when tooled into still-workable concrete. *The basic rule of thumb for the maximum distance between control joints on a 4-in.- to 5-in.-thick slab is 12 ft.* (For more information about joints for interior slabs see p. 43 and for joints as design elements see Chapter 3).

WE POURED UNCOLORED CONCRETE for this interior slab in the Natoma Street penthouse and then troweled in beach glass, turquoise, and mother-of-pearl. It was ground and polished and sealed with a penetrating sealer.

THE CURE

Curing refers to the chemical process of hydration, which results in the hardening of concrete. Proper curing translates to a durable surface and minimum cracking. Ironically, too much water in the mix and then too little water during curing weaken the concrete. Ideal conditions include high humidity (to retard water loss) and a concrete temperature between 70° F and 75° F during the first three to seven days after the pour (55° F to 85° F is acceptable). This is when temperature control is most critical. An exterior slab should be kept moist for at least three days—seven is preferred. Protect your fresh slab from rain for the first 24 hours. After the initial hard set, rain won't harm it. If it's 55° F or below, use heated water to mist the slab or build a tent over the slab and use a "salamander" propane heater to warm it. If it's over 85° F, shade the slab. Here are some methods for keeping the concrete moist as it cures:

- Misting or fogging for the first four or five hours, then kept damp three to four days.
- Curing paper, rolled directly onto the fresh concrete to prevent evaporation.
- If not planning on applying acid stain, a water-based curing compound or masonry sealer that creates a chemical film.
- Old rugs, sawdust, straw, burlap, or sand, which are periodically wetted. Don't use if the concrete has been colored or is to be stained.

The Basic Interior Slab

Most of our floors are interior slabs in new construction, poured over a wooden subfloor. Such floors present a few challenges, such as getting the concrete to the site, making sure the subfloor can take the weight, or finding the proper thickness of the slab so it will work with adjacent floor surfaces. As for pouring and finishing any concrete floor, the process is the same.

Designing a floor is a different challenge. Here are some things to consider in planning an interior slab.

THICKNESS AND MASS

Unless constrained by some design or structural consideration, we prefer to pour a 2½-in.-thick concrete slab on a wood-frame subfloor, as in the Natoma Street project. This thickness is a workable compromise between the greater weight of a thicker floor and the increased tendency of thinner slabs to crack. A 2½-in.-thick slab allows room for sufficient reinforcing materials and, if specified, the piping for a hydronic (hot water) radiant heating system.

Any wood-frame subfloor built to code should in theory accommodate the weight of a 2½-in.-thick slab (about 30 lb. per square foot) without deflection. *If you have any doubts about the integrity of the subfloor and its ability to support the concrete floor along with whatever goes on the floor—a grand piano, a heavy concrete hearth and fireplace surround—check with an engineer.* Deflection of the subfloor will probably result in cracks in the concrete, so some reinforcing of the subfloor may be needed. Or you may want to consider a thinner floor or lightweight aggregates.

>> **AS THE CONCRETE** is being poured, it's important that the crew avoids stepping on the tubing, reinforcements, or inlays that have been so carefully wired into place.

RADIANT INTERIOR SLAB ON FRAMING

With this type of slab, the framing must be sized to support the added weight of the concrete. Two layers of sheathing are sometimes installed, running in different directions. Welded wire "remesh" and ½-in. rebar provide reinforcement.

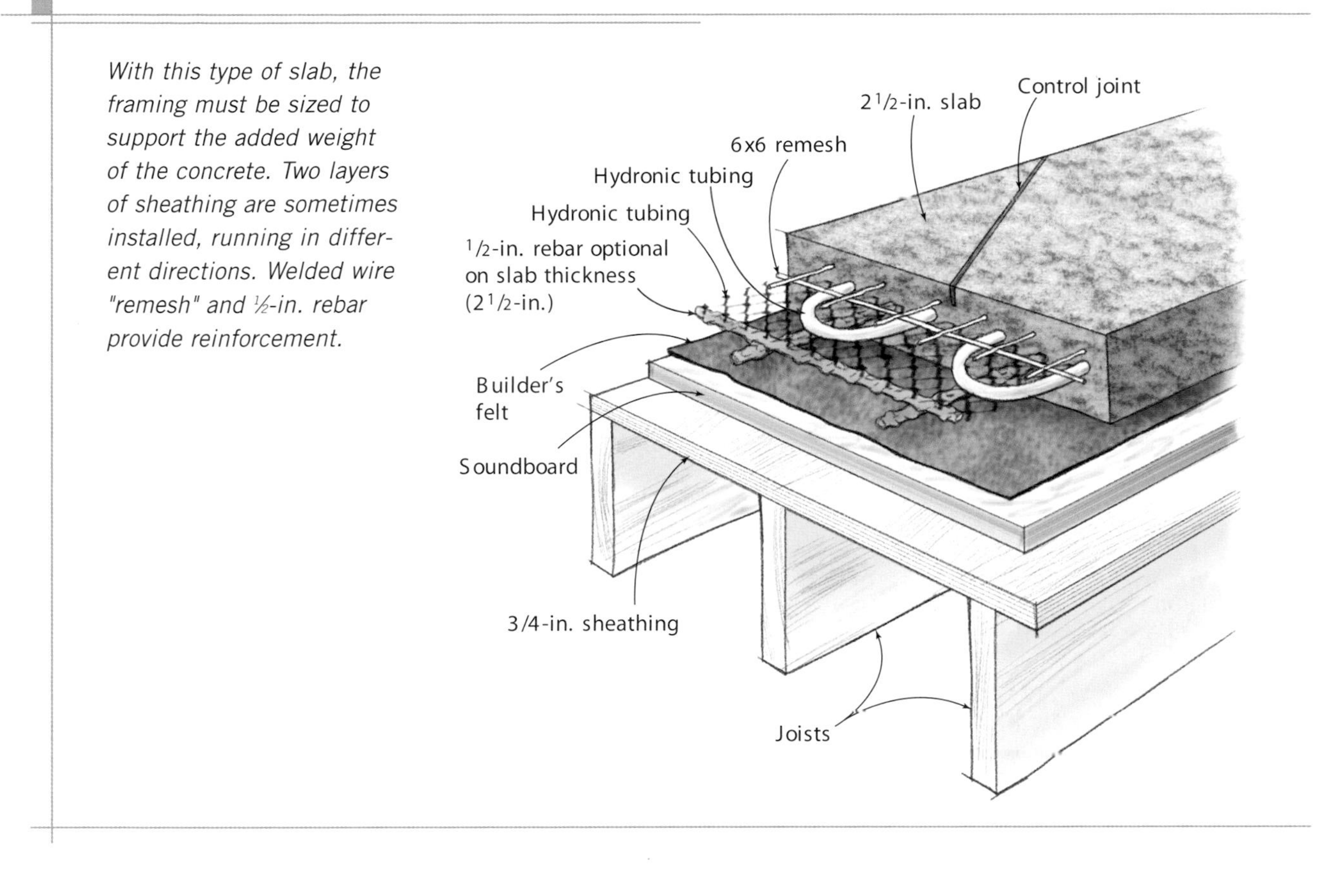

Thin floors—as thin as 1 in. (see p. 45 for information on compounds for very thin floors)—are typically specified in remodels to keep weight to a minimum or reduce level changes between floors. Thin floors are more likely to crack, thus good reinforcing is critical. Rebar is out of the question—within an inch of the surface, it can crack through or cause ghosting. We use expanded metal (or wire) lath and 4-in. wire remesh instead for floors thinner than 1½-in. Also of importance are the right mix and proper curing. A thin floor limits the depth of the inlays you can use; and if you're planning on a heated floor that's quite thin, an electric mat system may be the only alternative.

SETTING THE LEVEL

After the prep work is complete, it's important to make sure that the slab will be level. Use a laser or water level to mark the level of the slab along the walls and snap a line to the marks. This line shows where to place screed guides, and whether the floor is out-of-level.

An out-of-level subfloor shouldn't be an issue in new construction, but check it anyway because an out-of-level subfloor may result in uneven curing and cracks, or a problem called "potato-chipping," slight curling at thin edges. Watch for abrupt thickness changes, where concrete is more likely to crack. Feather these out with Fixall® or a self-leveling topping material such as Ardex® (see Resources, p. 208), a plastic polymer cement that doesn't shrink as it cures. Differences of an inch or even more, from one side of a room to the other spread across 10 or 15 ft., won't be a problem if the changes are gradual.

To keep a slab roughly the same thickness across an out-of-level subfloor, put down a layer of plywood at the lower, thicker edge of the floor and feather it out toward the thinner edge with Fixall or Ardex (don't use mortar, as it will chip and crumble at the thin edges). Expanded metal lath stapled to the existing floor can help "hold" the topping in place.

PLACING SCREED GUIDES

Screed guides are the rails along which the screed is pulled or "sawed." Since the guides establish the level of the finished floor, they need to be placed with care. For an interior floor in which the walls of the room form the perimeter of the slab, "hang" the guides on opposite walls. Guides are a 2x4 or 1x4 screwed temporarily to the wall with its bottom edge on the level line. A 2x4 screed is cut to fit with cleats at the ends that ride along the top of the guides during screeding.

For a screed, use a straight, stiff 2x4 or other board. A wooden screed shouldn't be longer than 10 ft., since a long screed will tend to sag or deflect during screeding (metal screeds can be longer).

THE SCREED GUIDES for this project are posts designed to accept 2x4 guides on adjustable hooks.

BY DESIGN

ON ONE JOB WE ISOLATED columns with a ring of sheet metal—interrupted by three inlayed stones— around each column base, which held back the concrete, creating a void. We poured the slab up to and around the band. Later, we pulled out the band, filled the void with gravel, and then cut in control joints radiating out.

For big floors, place one or more intermediate screed guides. In the living room of the Natoma Street project, a 30-ft. by 30-ft. space, we set temporary 2x4 guides mounted on special adjustable posts.

CONTROLLING CRACKING

Concrete made with an overly wet mix will shrink and crack more than concrete made with a stiff mix, and thin slabs are more likely than thick ones to crack. While control joints tooled into wet concrete, or cut in post-cure prevent cracks, proper underlayment is the first step in controlling cracks. We've found certain strategies work best for us.

THERE ARE PLENTY of control joints in this living room floor, but like the floor at Ahwahnee, this one cracked the way the concrete wanted.

AS THE CONCRETE IS SCREEDED, we pull out the guides and the stands so the voids can be filled in while the concrete is still wet.

PAPER AND PADDING

Use a membrane—roofing felt, plastic, foam padding—to separate the concrete floor from the subfloor, from surrounding wall surfaces, and from any pipes or ducts that extend through the floor. This helps prevent cracks that might form when stresses in the underlying structure are transferred to the concrete slab. What you use will depend on the substrate you're dealing with. For concrete on wood, roofing felt acts as a slip sheet to help prevent cracking during the curing process and to keep water from leaching out of the concrete while it's curing. For concrete on concrete, an isolation membrane is also called for when you're putting a new concrete floor over a concrete subfloor that is

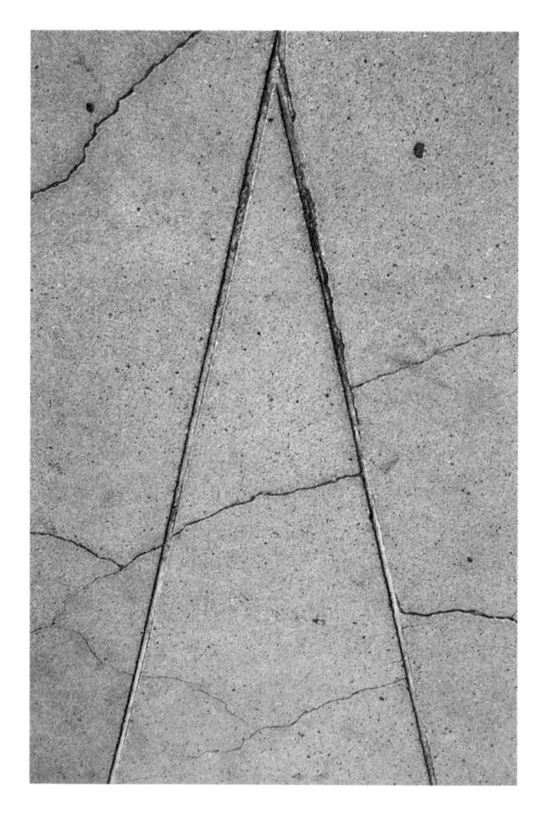

CONTROL JOINTS can only prevent so much cracking, and in this case, the elaborately designed Ahwahnee Hotel floor is enhanced by the cracks that come with age.

THE INTRICATE JOINTING of the floor in this cavernous sitting room of the Ahwahnee Hotel in Yosemite, Calif., has survived decades of wear and tear.

Setting an Inlay

The major drawback in placing inlays before pouring the concrete is that the inlays tend to get in the way while you're working. For large or heavy inlays, start by making a mortar bed. This mortar can be a mixture of sand and cement, Fixall, or a rapid-setting grout if you're in a hurry. Place the inlay in the mortar. Using a level or laser level, set the inlay to the established screed level.

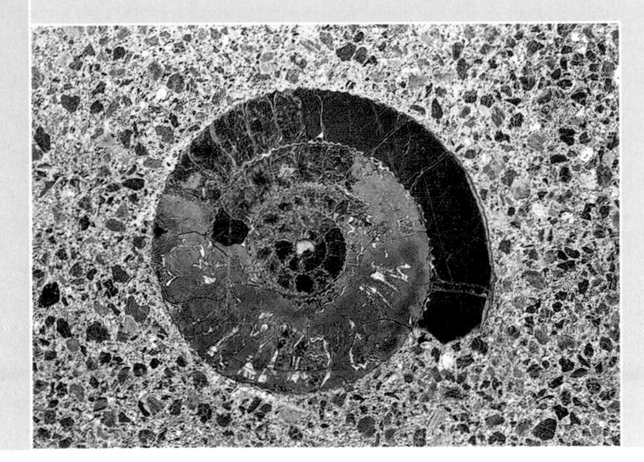

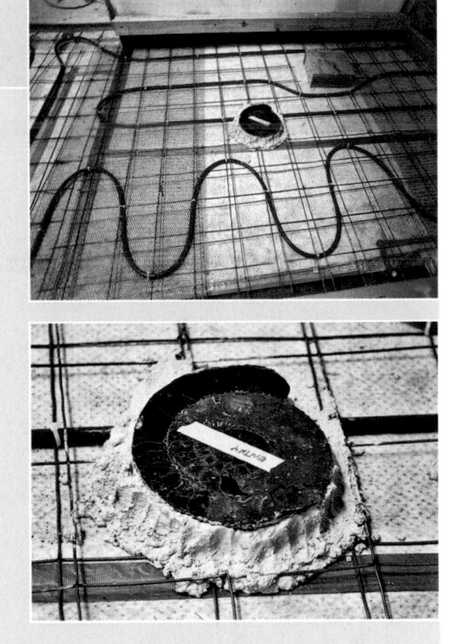

Because it's easier to place reinforcement like remesh before the inlays, cut out areas of remesh for the inlays, or place the inlays on it. To free the remesh so it can be lifted during the pour, cut out around the inlay. If you've placed the wet mortar bed on the remesh, either cut the remesh free of the mortar bed so it can be lifted during the pour, or gently lift the mesh up in the mortar bed to the desired level (no closer to the surface than 1 in.). Check the level of the inlay again and adjust if necessary.

badly cracked. If you're pouring over a structurally *sound* concrete subfloor, however, bond the new concrete to the old—especially if the new floor is a thin one. Bonding the two slabs together creates a single stronger unit. Apply a coating of an acrylic bonderizer to fuse the new slab to the old.

AROUND POSTS, AT CORNERS, ETC.

Pad outside corners where two walls meet by running ¼-in. packing foam or a similar material up the wall to the screed line. Also wrap pipes, column bases, and other interruptions. These cushions help absorb any stresses that build as the concrete shrinks.

Most of these precautions against cracking may seem like overkill to the seasoned pro, but remember that we're looking at the slab as a finished floor, not just an underlayment for another flooring material.

More About Thin Floors and Self-Leveling Compounds

Many products on the market allow you to pour a very thin floor (⅛ in. to ½ in.) over an existing concrete or wood subfloor. The products are useful for repair or covering an old floor that's badly worn, cracked, or stained. These overlays include Ardex, Americrete, and Aronite, among others (see Resources, p. 208.). Ardex, for example, is a self-leveling compound, meaning it will set up flat when it's poured onto an out-of-level substrate. (Others require troweling.) These products contain cement, fines, and an acrylic agent that creates a

» **CARBON FIBER MESH,** flexible and strong, is often used for thin floors.

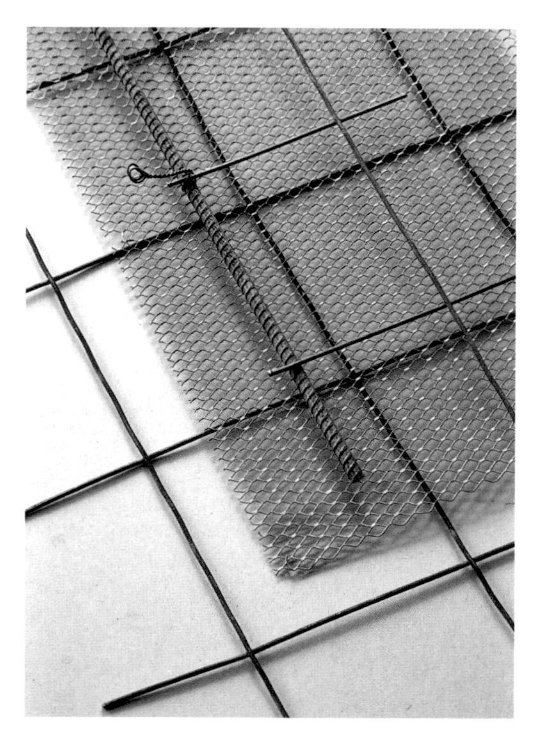

^ **FOR THIN CONCRETE SLABS,** remesh and expanded metal lath are good reinforcing options.

tight bond with the existing surface to reduce the risk of cracking or peeling. Because these overlays are so thin, they are typically reinforced with a fine cloth mesh or metal lath and require skill to apply properly.

REINFORCEMENT

Rebar can be used in floors as thin as 2½ in., but it's out of the question in very thin slabs, since the rebar will sit too close to the surface of the concrete. Other types of reinforcing materials are completely adequate and more suitable for thin slabs. These include:

- Pencil rod (very small-diameter rebar);
- Expanded metal lath: the type plasterers use, put down at intervals directly onto the subfloor (the combination of lath and another reinforcement gives a thin slab great tensile strength);

- Fibers: These come in a variety of forms, including nylon, polypropylene, fiberglass, and stainless steel. Some types of fibers are designed for structural purposes and aren't appropriate for concrete intended as a finished surface. We recommend nylon fibers. (see Resources, p. 208);

- Carbon fiber mesh: a very light, very strong, and very flat material ideal for thin slabs (see Resources, p. 208);

- Stucco lath and paper: the common "off-the-shelf" material used as an underlayment for stucco walls;

- Electric welded wire mesh (EWWM), a.k.a. remesh: wire reinforcement that comes in rolls or flat sheets.

On the Natoma Street project, we put down a layer of metal lath, fastened directly to the subfloor, and on top of that two layers of 4x4 remesh. The tubing for a hydronic heating system was sandwiched between the two layers of remesh. During the pour, we pulled the remesh and tubing up off the subfloor. Few cracks have formed.

SELF-LEVELING COMPOUNDS like the one used to top this floor can be stamped or acid-washed just like concrete.

FLOOR TRANSFORMATIONS:

3 Inlays, Linework, Stamps & Finishes

Following the design for a concrete floor requires more than simply hammering together some planks and pouring in the concrete. As a novice, I watched the concrete pros at work whenever possible. Then, on modest jobs of my own where my errors and omissions would be hidden under a deck or subfloor, I'd imitate what I had seen. I would then follow the standard procedures until I was comfortable, but soon enough, I was improvising from the basics—I couldn't resist adding a dash of Mexican pebble or a pinch of extra pigment, and years later I still can't.

As you'll hopefully discover, beautifully crafted concrete requires not only experience but a thoughtful approach at every stage of the process. There are many style techniques that you, too, can experiment with as I did (and do), processes that will personalize and transform your floor beyond a basic slab.

In the previous chapters, we discussed design fundamentals and the basics of pouring a floor. In this chapter, we'll cover in more depth techniques that belong in the repertoire of the traditional concrete contractor. You'll find the material contains a healthy respect for the discipline of the basics plus a twist of attitude from the artist.

« **THE DRAMATIC APPROACH** to this grand house is created by wavy alternating bands of broadcast Mexican pebble and salmon stone. The steps in the foreground are lightly textured with a broomed sand finish for safety.

THE COPPER BAR INLAY traverses both wood and concrete, "stitching" together the two materials, and acts as a border to the stained and unstained concrete.

We'll cover simple, cost-effective methods for enhancing the look of a floor after the concrete is poured (broadcasting color and stone) and while the concrete is still workable or firm but green (stamping, broom-and-wash finish, and other textures). We'll also consider several things you can do to enhance the look of cured concrete (grinding, acid-washing, polishing, dry-cutting). The techniques explained in the following pages are all challenging, but, with practice, ultimately within reach.

Jointing

Any type of joint can serve as a design element that might at least distract the eye from the inevitable: cracks. Lines—joints—can be tooled into the still-wet concrete or cut into green or cured concrete to separate sections, to create patterns on the slab, or to provide predetermined routes—control joints—for cracks to follow. The width and depth of the lines depend on function; control joints need to be deeper (no less than one-fourth the depth of the slab) than lines placed solely for design.

OUR CREW USED a long-handled jointing tool at the Clark Place patio to reach across the slab. The first pass pushed aside aggregates; later passes with a hand jointer cleaned the edges of the joint.

As noted previously, control joints are intended to deliberately weaken the concrete along the line of the joint. As stresses build in the concrete, any crack that forms will follow the joint, where it will be hidden.

Again, control joints are typically laid out in a grid. For a 4-in.- to 5-in.-thick slab, the recommended maximum distance between joints is 12 ft. Thinner slabs require more frequent joints. There's no reason joints can't be placed in a more imaginative pattern, but there are predictable lines of stress along which cracks will likely form: diagonally off an outside corner, for example.

Jointing the concrete when it's wet or still fresh is the best approach, since joints are placed before any stress cracks have had a chance to develop. Joints that are cut into hardened concrete with a diamond blade serve to control cracks that may form later.

Many passes with a jointing tool are required: The first pass takes place when the concrete is relatively soft in order to push large aggregates out of the way. As the concrete is repeatedly floated and troweled, the joints will be covered with cream, so repeated passes are necessary. Later passes take place when the concrete starts to firm up and becomes "shapeable." By this time there should be no large aggregates in the path of the tool, so that each pass smooths the contours of the groove.

IN THE HOGAN/MAYO RESIDENCE, all of the floors are concrete; the joint work was key to creating visual interest in otherwise monotonous planes.

POURING THE SLAB in separate stages means a more intricate jointing plan but translates into additional time with individual fields to create color and texture.

BY DESIGN

SOFT-CUTTING is a process we use to cut control joints into the hard but still green concrete. The cuts are made when the concrete is quite firm but before cracks have a chance to form—the day of the pour or the day after, depending on conditions. When timed correctly, the cut line has cleaner, sharper edges than a tool-in line. Cut too early, and the blade may dig out the rock, creating a jagged line. Cut too late, and the edges may chip. Experience rules here.

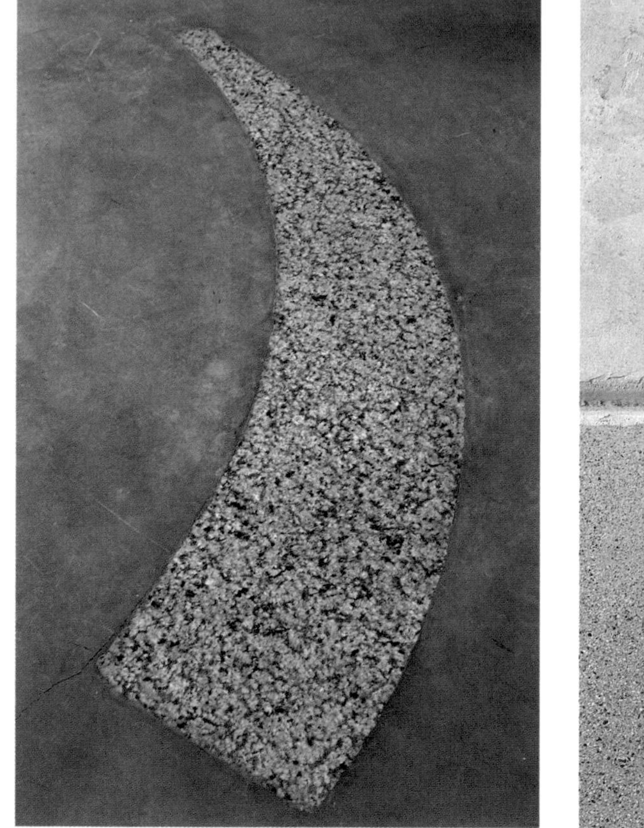

» **A FRAGMENT OF** an old granite countertop for this crescent-shaped inlay was placed in the concrete before the pour because of its weight. Otherwise, it may have sunk below the surface.

» **AT THE NATOMA STREET PENTHOUSE,** the copper bar, tooled on both sides, has an enhanced dimension in contrast to the plain tooled border line and ground concrete.

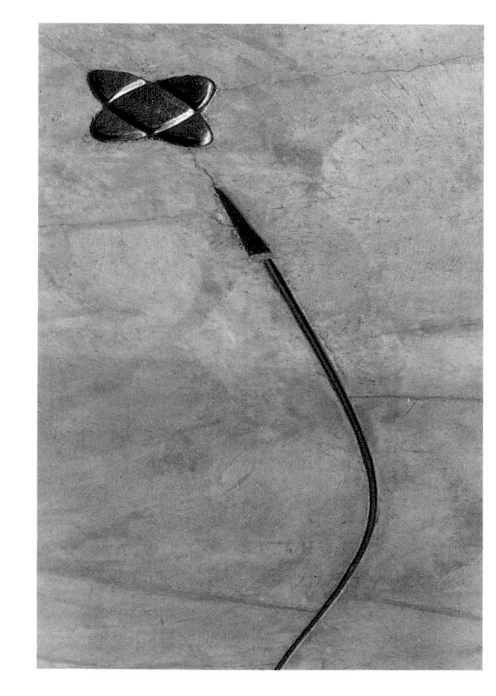

» **THIS INLAY WAS PLACED** before the pour on the McMillan project and made using a rubber mold formed around the handle of a water shut-off valve. The arrow itself is granite.

Inlays

There are two approaches to placing an inlay: One is to set the inlay down *before* the pour (see Chapter 2); the other is to put the inlay into the wet concrete during the pour. We do it both ways, sometimes in the same floor, and the decision about which way depends on a number of factors: How large is the inlay? Is it heavy and likely to sink in the wet concrete? Is it so small it could get troweled over and lost?

Inlays to be placed in still-wet concrete need to be fairly thin, typically no more than a couple of inches; otherwise, they'll be hard to place and might get hung up on the reinforcing materials or aggregates. Timing in their placement can be a delicate matter. If an inlay is large or heavy, for example, a piece of a car engine or a chunk of granite, there's a risk that it will sink. If you wait too long to place the inlay and the concrete has begun to harden, it could become difficult to push or pound any aggregates out of the way. Then you are forced to dig out a void for the inlay and smooth the concrete after putting it in place.

Inlays make troweling a smooth, flat floor more difficult, especially if they're not placed flat, or if they are bumped out of alignment during the pour. A good finisher can usually adjust, so any unevenness around an inlay will hardly be noticed. Grinding and polishing can help finish an inlay to a degree that is difficult to achieve just through troweling.

On the Clark Place slab, we placed two types of inlays in the wet concrete: a part from a car transmission and copper bar stock (¼ in. x ½ in.) laid at 2-ft. intervals. Reinforcing an inlay like common bar stock with an edging tool as we did

COPPER STRIPS WILL POP OUT of the concrete unless they are anchored. Here, small screws were drilled in at 16-in. intervals.

FOUND OBJECTS can bring balance and aesthetic appeal to any concrete piece.

AT THE CLARK PLACE PROJECT, the crew marked the sides of the form and then used a 2x4 as a guide when placing each copper strip.

WITH THE CONCRETE SOUPY, a light tap with a trowel was sufficient to settle each strip flush with the slab's surface.

creates a bold line with a slightly three-dimensional quality that enriches the look of the slab.

Each bar was tapped with a series of holes. Then we placed 1-in. screws in the holes to anchor the bars in the concrete. Copper will tarnish, but it won't corrode, and it's a good match for natural concrete. Traditionally, brass rather than copper is used for terrazzo, because brass works well with many different floor colors (most contemporary terrazzos are made with brightly colored epoxies). Copper bar is less expensive than brass.

«**AT THE CLARK PLACE SITE,** the first part of our project, the floor for the addition, is complete. The protruding rebar will support the planned concrete wall.

3 **ONCE ALL OF THE STRIPS** were in place, the crew went over the entire slab with a bull float to help level the strips and bring up more cream.

4 **THE CREW USED** a ⅜-in.-radius edging tool to create the rounded edge around each copper inlay.

5 **WHEN THE SURFACE HARDENED,** the crew scraped the thin layer of cream covering the rods and began to trowel the entire surface.

Integral and Broadcast Color

There are lots of ways to get "bang for the buck" with color for floors.

We sometimes use integral colors, powdered pigments that are mixed into the concrete before it's poured. This colors the entire batch of concrete. Integral coloring for a large batch of concrete for a floor can get expensive, though: for example, $4 per square foot for 8 percent loading of red for a 4-in. slab. Orange, ochre, and yellow pigments made with iron oxides are comparatively inexpensive, as is black. For a simple, "natural concrete," we generally add black to our mix to darken the concrete a bit, since uncolored concrete tends to cure to various shades of rather unappealing pale gray.

A color hardener (a.k.a. "dry-shake" hardener), is a finely ground mix of pigments, cement, and conditioning agents that produces a hard, colored surface. The hardener is broadcast onto the wet concrete and troweled into the "cream." This concentrates the pigments on the surface of the concrete, where they can be seen; a dry-shake hardener can produce a more intensely colored concrete than an integral color. (Color hardeners come in a variety of colors. Prices for commercial hardeners range widely, from about 20 cents a square foot for grays to more than $2 for the exotic blue-greens. See Resources, p. 208.) Commercial stampers use hardeners to create faux stone resembling flagstone, cobblestone, and brick.

1

FOR THE INTERIOR SLAB, the author has chosen a part from an automatic transmission for an inlay.

2

THE INLAY IS SEATED in the still-soft concrete.

3

NEXT, WE REMOVED THE PART, covered it with duct tape to keep the voids from filling, and put it back in place. The crew then troweled a paper-thin layer of cream over it.

On occasion, we've mixed our own version of a shake-on color hardener using pigment, cement, and a little sand. We've even scattered pure pigment using a flour sifter and then troweling it into the cream, but this isn't something we necessarily recommend. It's a bit risky. Remember, the maximum amount of pigment used, by weight, should be no more than 15 percent of the weight of the cement in the mix, since excessive pigments will weaken the concrete, leaving the floor vulnerable to wear. It's impossible to gauge the ratio of pigment to cement when you're simply scattering it on the surface.

LATER, AFTER THE CONCRETE CURED, we filled the voids in the inlay with a slurry. This, along with the rest of the slab, was lightly ground.

BY THE WAY

Be aware that concrete exposed to sun and weather will change color over time, and any coloring will tend to become muted. Concrete colored with ultramarine blue, blue-green, or with organic black pigments, whether integrally colored or colored with a dry-shake hardener, will be especially prone to fading. There are some synthetic blues and blacks designed for use in exterior concrete that will be less likely to fade *(see Resources, p. 208)*.

« **THE INTENSE BLUE-GREEN** color of this floor was created by broadcasting pure ultra-marine blue and green pigment followed by an acid-wash and color hardener, after the 28-day cure time

FOR THE PATIO, the next phase of our project, we applied a liberal amount of slate-gray color hardener to the wet surface of the fresh concrete. It was floated into the surface first with a wide, wooden float and then troweled in repeatedly through the remainder of the finishing process.

FOR THE CLARK PLACE PATIO, we placed 2x4s along the control joints to frame each field that received the pebbles and glass.

SCATTERING GLASS ON TOP of the pebbles adds depth and sparkle to the final look.

THOROUGHLY SATURATE the surface with the stones; as much as 50 percent will be "lost" to view.

Broadcast Stone

As is the case with so much of placing and finishing concrete, when you broadcast stone, timing is crucial and experience rules: Spread the stones too soon, before the concrete is firm enough, and they could simply sink (and given the cost of some decorative stone, this means the loss of no small investment). Spread the stones too late, you may not be able to work them into the concrete sufficiently or trowel up enough cream to hold them; the result will be an increased risk that some of the stones will slough off under foot traffic. Likewise, troweling over the stone requires a deft touch: Bring up too much water with excessive troweling and the paste locking in the stones may be too weak to hold them securely in place (remember, too much water in a mix produces weak concrete).

If you're looking for a uniform application of exposed stones, in other words, it's not a bad idea to hire a pro, at least for this part of the job.

On the Clark Place patio, we spread ¼ in. Mexican pebbles in alternating bands that corresponded to the pattern of copper strips on the adjacent slab. We started about an hour and fifteen minutes after the pour began, spreading the pebbles liberally. We followed with a scattering of dark blue washed glass on top of the pebbles. The pebbles and glass were then steel-troweled into the cream until they were completely buried just under the surface. We also scattered a small

BY THE WAY

Be sure to moisten the stones *before* you apply them to the wet concrete; this will help create a stronger bond between the stones and the concrete.

THE FIRST PASSES with the float brought up some of the cream. Later passes with a trowel covered the pebbles and glass.

WE CAME BACK the day after the pour and used a stiff wire brush to scrub away some of the cream, exposing pebbles and blue glass.

AS THE JOB PROGRESSES, the concrete becomes more difficult to work.

amount of washed glass onto the adjacent bands and troweled them into the cream, intending to expose them later. It was important to completely cover the pebbles and glass at this point, so that when they were exposed enough concrete remained to hold them in place.

Broadcast stones are typically exposed by washing the surface with a strong, fine spray of water when the concrete is still green but firm. This might be several hours or even a day after the pour. On the Clark Place patio, because of the alternating bands, washing in this way would compromise the areas that we planned to broom-finish. We decided to expose the stones by taking a stiff wire brush to the cream once the concrete had set up completely. The weather was quite cool and damp, and so we waited until the next day for the

THE FLECKED ENTRYWAY to this Palo Alto, Calif., home is the result of broadcast stone that's been ground and polished.

LIKE INLAYS, stamping can be created with found objects. This delicate pattern was made with an Indian batik wood stamp and a car spring.

desired hardness. The pebbles and glass were exposed by scrubbing the surface with the brush. This gave us the precision we needed to keep the adjacent broomed textures intact.

Stamping

As noted earlier, commercial stamping produces (and produces and produces) accurate imitations, but we prefer to let concrete speak for itself and tend to use stamps discreetly, in ways that bring out concrete's natural beauty. These stamps are typically found objects, a batik woodblock or part of an auto transmission, for example. Anything can be used to stamp the concrete as long as it's durable and doesn't have a tendency to stick and pull up the finished surface.

The way we use these found objects, stamping is pretty simple. We'll take our stamp, place it on the wet surface of the concrete, and tap on it a few times with a hammer to create a stamped impression.

For the Clark Place interior slab, we used a batik woodblock and a short strip of leftover copper. On the patio, we created our own rubber stamp using a found object, in this case a sheet-metal panel from an old ceiling. We incorporated the panel into a 2-in. deep mold, mixed up a two-part rubber compound, poured it into the mold, and a day later had a beautiful, workable rubber stamp.

BY DESIGN

BECAUSE CONCRETE is a medium that slowly evolves over time, you can incorporate this movement into your design. Outdoors, there's no reason not to scatter glass, pebbles, or fossils on wet concrete and cover them with a thin layer of cream, as we did on the Clark Place patio. Eventually, foot traffic and weathering will expose new patterns.

ON TOP OF A RETAINING WALL, we pushed a steel ring into the concrete, then pressed an ammonite into the center in order to create this detail.

THE SMALL STAMP, a classic lotus pattern, was pressed about ⅛ in. into the concrete.

HERE, WE'VE REPURPOSED another found object, an Indonesian batik print woodblock, into a design tool. We placed the block and tapped it lightly with a hammer to make the impression.

1 ON THE CLARK PLACE PATIO, a dark blue-gray release agent was spread liberally at the site to be stamped.

2 THE STAMP IS CAREFULLY POSITIONED so the release agent is not scattered to the adjacent concrete.

3 HERE, A CREW MEMBER STANDS on one edge of the stamp while he tamps the other. It's important to distribute weight evenly to produce a uniform impression.

Stamping D.I.Y.

Creating your own stamp tool is straight-forward and fun. We used a salvaged 2x2-ft. square from a tin ceiling for the decorative stamp on the Clark Place patio. The Art Deco–style star design had a shallow relief, so it would read well on concrete. The release agent added a gray-blue hue, which mixed with the moisture in the concrete and accentuated the detail nicely.

A MOLD WAS CAST from a tin ceiling panel. Here, release agent is sprayed into the mold.

THE MOLD COMPOUND comes in two parts. Working time after mixing is about 5 minutes.

THOROUGHLY BLEND the two-part urethane with a low-speed drill and paddle.

POUR EVENLY and methodically to prevent voids from trapped air.

AFTER A DAY'S CURING, the tough urethane stamp pulls cleanly from the tin mold with release agent.

« **ONCE THE SLAB HAS THOROUGHLY CURED,** green acid stain and penetrating sealer will be applied.

Stamping with found objects entails some risk and should be approached in a sense of playfulness. Commercial stamps are designed to consistently pull free from the concrete, leaving behind a clean impression. A found object, such as a batik woodblock or auto transmission part, on the other hand, isn't intended for this sort of use. There have been times, usually when we've tried to stamp when the concrete hasn't set up sufficiently, that we get a ragged hole instead of a clean impression when some of the concrete pulls up with the stamp. But it's easy enough to fill in the hole and try again. And there have been times when we've tried to stamp into concrete that's gotten a bit too hard, and we've ended up with no clear impression and a broken stamp. If you're new to concrete, we recommend making a few small samples and experimenting with stamps, to get a feel for the timing.

Commercial stampers, because they make so many impressions, use release agents so that the stamp pulls away cleanly. Some commercial stamps rely on a thin plastic sheet in place of a release agent, which serves the same purpose. Very often, the release agent is colored, and it acts like a color hardener; it's spread on the wet concrete, then is driven into the cream on the surface as the impression is made. A release agent doesn't get troweled into the concrete, though, and most of it will wash

A COMMERCIAL STAMP is perfect for making basic improvements to a plain slab on grade, like this patio which looks like brickwork.

away unless it's sealed with some type of topical sealer. We used a release agent with the rubber stamp on the Clark Place patio and left a residue to accentuate the stamp relief.

Whether you use a found object for a stamp, a homemade stamp, or a commercial stamp, there are a few basic things to consider:

- No stamp cuts deeply enough for the pattern to serve in any way as a control joint, so a stamped slab, like any slab, will need joints (joints and stamped designs will need to be integrated). Typically, the control joints can serve as dividing lines between stamped fields, and may demarcate pattern changes. Take into account the width of the control joints when laying out your stamped patterns.
- Do your layout *before* the concrete arrives, so you know exactly where you'll start and finish; you don't want to be figuring this out while the concrete is setting up.
- The best concrete mix for a stamped floor has 6 to 6.5 sacks of concrete per cubic yard, and ⅜-in. pea gravel. Large rock may make it more difficult to stamp to the desired depth/relief.
- Tell the batch plant that you intend to stamp the concrete, and find out if retarder or accelerator is recommended for the prevailing weather conditions in your area.

THE FINISHED STAMP brings detail, dimension, and color to the Clark Place patio.

<< **THE ENTRYWAY IS EXPOSED** to the elements and often wet from sprinklers, so a salt and stain finish was applied.

<< **THE CONCRETE SURROUNDING** this inlay, a boot scraper, has a washed finish, creating a non-slip surface that appears matte.

Broom-and-Wash Finishing

After being steel-troweled for the last time, when the slab is nearly set, the surface can be swept with a horsehair broom, leaving directional lines in the slab. This is the classic way public sidewalks are finished to give the surface some "tooth" so they don't become a slip hazard. A wash finish is done when the concrete is harder but still green. Some of the cream is washed away with a strong, even spray of water. This leaves the fine sand in the mix exposed and gives the slab a non-slip surface.

Grinding and Polishing

Hard as a cured concrete floor may be, there's plenty you can do to it to alter, refine, or enliven its surface. Indeed, in many ways a cured concrete floor can be treated very much like a blank canvas.

<< **ON THE CLARK PLACE PATIO,** we broomed the strips without broadcast stone and glass, cleaning the brush frequently so concrete in the bristles wouldn't leave gouges in the surface.

BY THE WAY

If you grind and polish the surface of concrete pigmented with a color hardener, it's possible to grind away some of the color.

One advantage of integral color over a color hardener is that the concrete is colored throughout, so lots of grinding won't affect the color, or at least not by removing the pigmented surface. Grinding down far enough to expose aggregate *will* affect the color whether you've used an integral color or not: A concrete's color tends to shift toward the color of aggregates—typically brownish or slate-gray—as they're exposed.

The technology for grinding and polishing concrete has advanced significantly in recent years, making it possible to easily put a glass-like finish on a concrete floor. A highly polished floor has many advantages over a floor with a more textured surface. A polished floor requires comparatively less maintenance than an unpolished floor. In a commercial setting a polished floor provides a very durable surface on which it's easier to shift merchandise, shelving, or equipment. Many new warehouses have diamond-ground and polished floors—the flat surface means less wear and tear on fork-lifts.

The best machines for concrete grinding and polishing are based on stone-polishing technologies and have a "planetary drive." This type of drive has one large disc that rotates in one direction and

>> **THIS SMALL GRINDING MACHINE** uses steel-backed diamond pads.

ON THE CLARK PLACE SLAB, Jim Lundy (see Resources, p. 208) used a coarse 60-grit pad to expose the aggregates in alternating bands.

smaller "satellite" grinding discs of diamonds embedded in metal or resin, which rotate in the other direction. This planetary system creates a random grinding pattern that lowers the risk of scratches while producing an extremely flat surface. These machines come in a variety of sizes, and typically weigh 300 lb. or more (the heavier the machine, the more aggressively it will grind the concrete).

Grinding and polishing can done wet or dry—few agree on which approach is best. Most grinding and polishing machines made for wet-grinding have a built-in water feed. The water cools the pads so they last longer and keeps the dust down. But wet-grinding creates a huge mess, especially if you're taking off a lot of concrete. (The sludge produced by grinding and polishing is vacuumed up during grinding.) Dry-polishing using a machine equipped with a dust collector eliminates the problem of sludge. Some contractors will begin by grinding dry, when they're removing quite a bit of material, and then they switch to wet-grinding for the final polishing phases, when less residue is produced.

« **IF YOU LOOK CLOSELY** at the area surrounding the machine, you can see the already-ground concrete, which is darker and more textured.

BY DESIGN

IF YOU ARE PLANNING to expose the aggregates by grinding, there are some things you can do when mixing and pouring to make it easier:

- Seed the concrete with decorative pebbles, glass, shells, or even metal scraps; trowel over thoroughly as with broadcast stone, then grind and polish to expose.
- Specify a low-slump mix, so the heavier aggregates will be less likely to sink.
- When placing the concrete, try not to walk across a very wet pad, as this can push down the aggregates. The result with grinding can be a "ghost" of footprints where there's less exposed rock. If you must walk across the wet concrete, stir each footprint with a shovel or trowel to bring some of the aggregate back to the surface.

The key to successful grinding and polishing is the diamond pads. Pads for fine polishing consist of diamonds embedded in a somewhat flexible resin matrix. Pads for fast, aggressive grinding have a metal base on which the diamonds are affixed through electroplating. If you're planning to take off a lot of the top layer of cream to expose the aggregates, you'll begin with a very coarse pad, say a metal-backed 40-, 50-, or 60-grit pad. From there, you'll progress with increasingly finer pads (from a 150-grit to resin-backed 300-, 400-, 800-, and 1500-grit pads). For a very glossy finish, the final polishing may be with a pad as fine as 3000 grit, or you may want to spread polishing compound on the floor, for even more sheen. Or you may wish to stop at 400 to 800 grit, leaving the floor with a little tooth. For aesthetic or practical reasons, you may not want a very slick floor. But sealing, waxing, and buffing concrete that's been finished at 400 to 800 grit makes an attractive floor.

Acid-staining

Acid-staining has enjoyed a renaissance since its first appearance in the 1920s. It's a simple, relatively inexpensive way to permanently color concrete or create patterns and images. The range of color is limited to tans and rusts, greens and blue-greens, and shades of black. Due to the nature of the staining process, acid-stained floors are usually mottled. It's impossible not to get variegation and unexpected results—this is both the charm and bane of the process. Experiment on that old garage slab before attempting to tackle the living room floor.

1 **A SLAB DOESN'T LOOK LIKE MUCH** before any staining has been done.

2 **CUTS ARE PLOTTED** with precision before they are made in the completely cured slab.

3 **AFTER DRY-CUTTING AND ACID WASHING,** the floor of the reception area in this small business has graphic balance and color full of depth and dimension.

HERE IN THE DINING ROOM of the Ahwahnee Hotel in Yosemite, Calif., the intricate intersection of dry-cuts and acid-washed fields of concrete are a sharp design contrast to the precisely-executed linoleum inlay work in the main hall.

CURVING COPPER STRIPS and mica chip inlays add to the rich tone produced by acid-staining on this bathroom floor in the Hogan-Mayo project.

IN THE SEBASTOPOL, CALIF., KITCHEN, the floor has been divided by transecting dry cuts, then stained in sections, giving it a softer, slightly aged look.

ON THE CLARK PLACE SLAB in El Cerrito, Calif., a liberal coating of green stain was applied (1) with a spray applicator. The surface was then brushed (2) so the stain could penetrate and react with the concrete (3).

An acid stain consists of a water-based solution of hydrochloric acid and metallic salt pigments. The acid etches or opens the surface of the concrete, which allows the salts in the stain to penetrate below the surface, where they react with the free calcium hydroxide in the concrete, forming permanent bonds.

The stain can be sprayed or brushed on, and it can be used like paint to create detailed, intricate designs. These acid solutions can be diluted with water to vary the shading and can be used with concrete dyes.

PROPER PREPARATION

Because an acid stain is translucent, any oil stains, paint, framers' marks, glue or mastic, or blotches of old paint, plaster, or joint compound on the underlying concrete may show through.

« **THE RESULT IS A GREEN COLOR** that nicely accents the slab, which will be a pleasing focal point for the room when the addition is finished.

BY THE WAY

Acid stains are corrosive and can cause serious injuries to skin or eyes, or to the lungs if inhaled. When you're working around stains, there are some basic precautions to take:

- **Be sure there is adequate ventilation. Open all windows and doors if you're working inside.**
- **Wear a respirator that filters hydrogen chloride.**
- **Wear protective, splash-resistant goggles.**
- **Wear acid-resistant gloves, a long-sleeved shirt, long pants, and acid-resistant boots.**
- **Acid stains contain heavy metals, which are extremely toxic. Clean up with a wet-dry vac and dispose of residue at your local hazardous waste facility. Do not wash out into the soil.**

Sealing and Waxing a Floor

For interior slabs, you have the option of acrylic, urethane, and epoxy sealers. Acrylics breathe and they're easy to apply, but they're not as tough as other sealers. Urethanes are more scratch resistant, and epoxies are the toughest of all, but they're also more difficult to apply. The best sealers for an exterior slab will be solvent-based acrylic sealers that don't trap moisture in the slab. A lacquer-based penetrating sealer, whether used indoors or out, will give the slab a wet look.

Application of any sealer is fairly simple. First, make the sure the floor is completely cured, dry, and clean before you apply any sealer. Also, if you applied an acid wash, be sure all residue is cleaned off. The acrylic or urethane sealers can be rolled on with a lint-free roller with a 3⁄16- to 1⁄4-in. nap. Epoxy sealers, which are thicker, can be applied with a notched squeegee, then rolled. After the sealer has dried, the floor can be finished with several applications of a mop-on floor wax.

Oil may come off with a degreaser. Paint, drywall, or joint compound can be scraped off. Chemical stripping agents may work to remove any mastic that had been used to place tile or carpet. Or scrub the floor with a solution of Tri Sodium Phosphate, or TSP (one cup of TSP in a gallon of water).

Mask areas to be left unstained, or stained with a different color, with blue painter's tape and masking plastic (most convenient is the type that comes already attached to the tape).

APPLYING STAIN

Acid stain is most often applied with a hand-pump sprayer with a conical spray tip onto a wetted floor (this helps prevent abrupt overlap marks and splotching). The sprayer shouldn't have any metal

THE MAKING OF THIS FLOOR by Kelley Burnham and Dana Boyer, for a private residence, began with a gray concrete slab.

MICHAEL MILLER AND KAREN TIERNEY apply many treatments to their floors. This one, a work in progress, received a light salt finish, was then stained buff and broadcast with ironite fertilizer. Here, we see "butter cut," a type of self-adhesive stenciling material being applied before sandblasting.

parts that might come in contact with the acid. A plastic garden sprayer works perfectly. Wet mop or brush the stain on for variegated effects.

FINISHING UP

Leave the stain on the concrete at least six hours, time enough for the pigments and cement to react completely. (You can remove the stain sooner, if there's a certain effect you're after. You may need to leave it on longer if the weather's cool and wet.) The residue that's left can be scrubbed off with water and a stiff brush, or a rotary floor scrubbing machine for large spaces. Use a wet vacuum to clean up the sludge as you work. Be careful not to track the residue onto unstained portions of the floor or areas with a different color; the footprints may be hard to clean off.

BEYOND ACID STAINS

Design techniques like acid-staining have inspired artists to take the process beyond conventional commercial applications. For example, concrete artisans like Michael Miller and Karen Tierney of The Concretists, Dana Boyer of ConcretiZen, and Gerald Taylor treat the concrete as if it were a

BY DESIGN

INTERESTING VARIATIONS in shading can be achieved by varying the amount of troweling across the slab. Each pass with a trowel, as the concrete sets up, creates a harder, smoother surface that won't accept the stain as well as more lightly troweled areas. The result will be random patterns of slightly lighter and darker coloring.

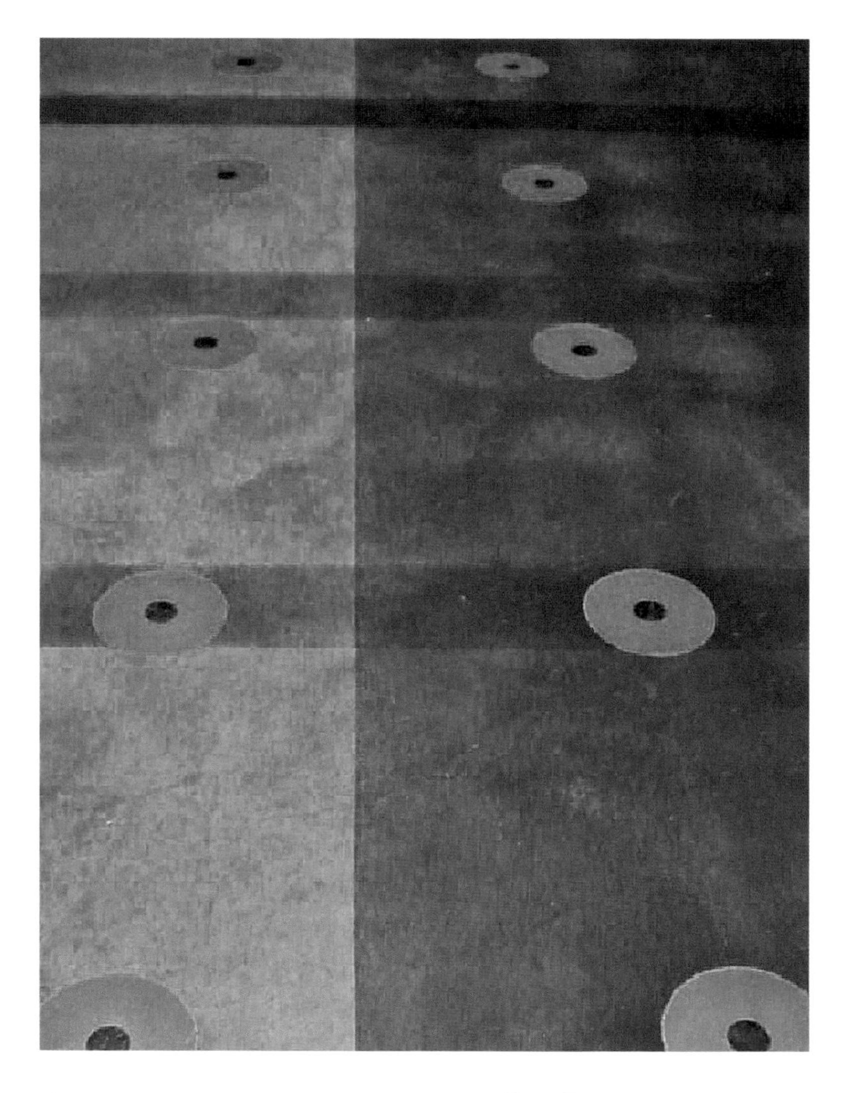

STRIKING COLOR COMBINATIONS suffuse The Concretists' work. Here, a floor ironically called "Ceiling Reflection" shades from orange to burnt sienna and features green "lights."

blank canvas, ready to receive complex, detailed images. They use a variety of materials and techniques to achieve their effects. Some of their tools and techniques include:

Concrete dyes and other coloring agents. Unlike stains, dyes don't react chemically with the concrete. Rather, they penetrate into the concrete. Dyes can be mixed, like paints, to create an unlimited palette of colors and tones. They're applied like stains. Pastels, charcoal, and graphite can be used to color or draw on concrete, but they depend on an epoxy or urethane sealer to keep them from wearing or washing off.

Sandblasting and engraving. Sandblasting, particularly with stencils, or engraving after staining can produce intricate designs in the concrete. There are several manufacturers that offer stencils with stock designs. Or you can cut your own from rubberized, adhesive stencil material such as Scotch™ brand No. 507 "Buttercut" (see Resources, p. 208).

Resists. A resist is anything that's laid, poured, brushed, or otherwise applied to concrete that prevents stain or dye from penetrating. A concrete sealer is commonly used as a resist, applied before the stain, leaving an unstained area on the finished floor.

Fertilizer. Some fertilizers contain the same mineral salts as stains. One, Ironite®, consists of small iron pellets that can be scattered on the surface of wet stain, leaving brownish patterns on the floor. Muriatic acid in powder form can be scattered to create feathery patterns that tend to be blue-green in hue.

«**THE FINISHED FLOOR,** sealed with a solvent-based acrylic to protect the finish, complements the adobe-style interior.

AFTER STAIN, dye-washed crayon, and crushed pastel were applied, the floor was air-brushed with iridescent metal powders.

For each of the methods and techniques we have covered in this chapter, there are professionals who can skillfully perform the task. Our purpose is to make theses processes accessible so that you understand the dimensions of both design and personal expression that go into the mix.

BY THE WAY

Harder concrete is easier to polish (it's also easier on diamond pads, which, somewhat counterintuitively, are worn out more quickly on soft concrete). Some contractors use a chemical hardener such as sodium silicate, lithium silicate, or floro-silicate. The hardener is brushed onto the concrete, typically after grinding and before polishing with the fine 800- or 1500-grit pads. The hardener reacts with the calcium hydroxide in the concrete and the result is a toughened surface. If you have the time, grind the floor when the concrete is still green, say, at 10 days, then wait another 10 days or longer to begin polishing.

4 NEW COUNTERTOPS: Classes, Products, Tools & Techniques

My first book, ***Concrete Countertops:*** *Design, Forms, and Finishes for the New Kitchen and Bath,* seems to have come along at just the right time. Homeowners, designers, and builders had apparently seen enough of granite, marble, and synthetic countertops and wanted something different. Concrete opened up a world of creative possibilities.

I've heard from many of our inspired readers who are now working creatively in concrete, some for the first time. Many have sent us examples of their work, and I'm pleased to include some of these in a Reader's Gallery (see p. 80). These readers responded to the accessibility of concrete and the fact that such a common and familiar material, one that had long been disparaged as an unsuitable finish material for the home, can be so energizing to work with, and look and feel so beautiful.

Many contractors and designers who already work with concrete have begun adding countertops to their repertoires, inspired by the possibilities as well as by consumer demand. An entirely new industry devoted to the design and fabrication of concrete countertops and other pieces has sprung up. But the surge in interest in concrete as a finished material is not only creating new business and design opportunities, it's presenting some interesting challenges. Although the tools and hardware for

« **AT THE TÉANCE/CELADON TEA ROOM** in Albany, Calif., the size, shape, and details of the tea bar have all been refined in the service of tea. Pieces of porcelain, crackleware, ammonite fossils, jadite and turquoise fleck its polished surface.

FOR THIS POWDER-ROOM VANITY in the Natoma Street penthouse in San Francisco, wood and concrete are a fusion of interlocking materials, underlining the strength and craftsmanship of the piece.

fabrication of finished concrete have made the technical end easier, the subtle "software" side of the equation—good design—remains as elusive as ever.

Traditional, Transitional, and Contemporary Styles

In the first book, we focused on countertops and the surrounding features. They were fabricated for projects that we at Cheng Design had designed in full, and they were modern with a natural, warm side. Yet perhaps as many as 90 percent of homes

THE INTERLOCKING PIECES of the countertop are reminiscent of Japanese joinery.

BECAUSE THE COUNTERTOP is largely a dark solid form, the details create both a pleasing distraction and a transition from one section of the countertop to another in this kitchen by Paul Discoe of Joinery Structures.

The Reader's Gallery

A number of readers of our first book sent us photos of their work, and you'll find here several examples that we've selected for innovation, good design, or professional execution. Please turn to the Resources section beginning on p. 208 for contact information.

COUNTERCAST DESIGNS in British Columbia casts many sinks like this shallow one that was precase in an MDF mold and integrally colored with red oxide pigment.

JAMES MCGIRE AND ANDREW SIMON, concrete designers in Hawaii, have cast this tricolor piece as a clever color sample. It was made with separate batches colored with wine, jade, and indigo NeoMix®. Each batch was simply poured into the mold, but not mixed together.

MARK FOREMAN'S CURVED COUNTERTOP for a steakhouse was cast in place. The surface was lightly ground after it was troweled. The copper was applied using a mixture of epoxy and copper, which was labor-intensive but could be carried out on site.

THIS 30-FT. BAR COUNTER, created by Michael Karmody of Stone Soup Concrete, at the La Cazuela Mexican Restaurant and Cantina in North Hampton, Mass., illustrates a pleasing geometry of design, where the large inlay, a circle, joins together the two long counters. The other inlays are fused glass with lights mounted under them.

DALE BLAYONE of Concrete Cuisine has crafted a unique kitchen countertop of falling elliptical tiers. To define the shapes, he used Autocad®, a computer design system, to produce templates, which he traced onto the melamine forms.

FOR THE HOLIDAYS, the client dropped in some Christmas ornaments where a Zen sand garden was originally placed.

today have what might be called "traditional" kitchens and baths—the cabinetry and other features are evocative of other eras, whether Craftsman, Classical, French country, English manor, etc. So the question is: How do we fit concrete honestly into an artificial traditional setting without resorting to imitating the imitation? Here are a few suggestions, derived from our own experiences wrestling with this very question.

DETAILS, DETAILS, DETAILS

Since the cabinetry in a kitchen or bathroom will normally inform the design, often a small detail from the cabinets—a subtle bevel or a bull nose with a little ogee detail, perhaps—worked into the concrete can allude to classical French or English wood detailing, tying the countertop into the rest of the room. Such simple but evocative details can be crafted out of wood or rubber in the mold. A bit of mosaic tile along the front edge of a countertop or inlaid along a back splash, can achieve the same end. In a Santa Fe or adobe-style home, something as simple as a piece of traditional Mexican tile can render the countertop relevant to that setting; some blue and white tile can pick up a Provençal look.

BY DESIGN

BUT MUST IT MATCH? Must one pick up on the style of the surroundings? Sometimes the countertop can exist just as it is; it doesn't have to slavishly adhere to the dominant style of the room. Consider all the stainless-steel industrial-strength stoves that now grace many a high-end kitchen in seeming disregard of the surrounding style. Few question this particular stylistic anomaly.

» THE CONCRETE COUNTERTOP features a slate mosaic drainboard area that was placed in the mold and cast integrally with the top.

THE COUNTERTOP INTEGRATES a perforated stainless-steel removable soap dish and sink along with a slate inlay.

SEPARATIONS

Another way to place a contemporary countertop into a traditional setting is to create a small separation between the countertop and the cabinets—a reveal or an intervening material might do the trick. For example, sometimes we'll run a strip of zinc or stainless steel along a wooden base between a countertop and the cabinets it sits on.

CRAFTSMANSHIP UNIFIES

We think of *craftsmanship* as a hallmark of earlier design traditions, but good contemporary design demands a level of craftsmanship at least equal to that practiced in earlier eras. Contemporary design relies on materials, colors, and forms rather

than on moldings and decorations for its effects. This lack of adornment and the simplicity of line often demand a high level of planning and craftsmanship.

And it is in this area of craftsmanship that contemporary and traditional design find common ground, and where concrete fits in so well. Mold the concrete around a wall or a post, integrate it neatly with a wooden, glass, or steel piece, and the

CONCRETE TRANSFORMS THIS KITCHEN on a modest budget into an elegant jewel. Mounted on off-the-shelf modular cabinets, the countertop was designed to maximize the impact of the dramatic island without using expensive accessories.

result is joinery that alludes to the joinery of, say, the wood wainscoting of a Craftsman-style room. People will respond to the unifying effect of the craftsmanship, even if they're not sure why the concrete fits into the setting so well.

CRAFTSMANSHIP ON A BUDGET

But craftsmanship need not be absurdly expensive. The Tsuchiyama kitchen, for example, was done on a relatively modest budget. Here, design and a strategic use of fine craftsmanship create a rich look balanced with quite ordinary materials. The cabinets are off-the-shelf, store-bought modulars, well made but inexpensive. The dark island wall is clad with black Formica; the countertop and backsplash in the background are simple black granite slabs. Only the concrete countertop itself is a custom "splurge." In all, the kitchen is an example of how a sculpted concrete counter can elevate an off-the-shelf modular kitchen above its humble origins.

HERE, INLAYS OF GRANITE meander toward the edge of the sink.

ADD SACKED CONCRETE and pour—our one-step proprietary countertop mix formula, NeoMix.

NEOMIX SAMPLES, like paint chips, show the range of available color and the progressive gradations of color brought out by grinding.

A COUNTERTOP SEMINAR in session. Contractors and concrete specialists rub elbows with architects, designers, and homeowners to learn the latest techniques.

FOR A BASIC SEMINAR at Cheng Concrete Exchange Academy, there is plenty of hands-on experience as witnessed by this pile of "homework."

New Developments

The rapid growth of interest in concrete as a fine finished material has led to a variety of new products, technologies, services, and even periodicals devoted to concrete as a finish material, such as *Concrete Decor* and *Residential Concrete*.

CLASSES

At Cheng Concrete Exchange Academy, we've begun offering one- and five-day intensive seminars—beginning and advanced—in the design and fabrication of countertops, walls, and fireplaces, and several other organizations such as the Portland Cement Association are beginning to respond to the growing interest with similar programs in decorative concrete (see Resources, p. 208).

COUNTERTOPS SHOULD BE SHIPPED on edge, like glass, and require foam-lined crates with built-in pallets.

CONVENIENT HANDHELD diamond pads for grinding, polishing, and refinishing come in a range of grits.

ONE-STOP SHOPPING

Internet sites, such as *concreteexchange.com*, provide information about techniques, services, and products for concrete pros and do-it-yourselfers alike. We've developed NeoMix® Pro-Formula, a proprietary blend of pigments and admixtures that, added to ordinary sacked concrete, makes countertop pouring a one-step process. Online, decorative aggregates, polishing pads, off-the-shelf sink and faucet knockouts, and slurry kits to fill voids are all available—everything you'd need to craft and finish a fine countertop (see Resources, p. 208).

PRODUCTS

As the popularity of concrete in the home has spread and the number of contractors crafting and installing decorative concrete has increased, the market has introduced a number of new products to meet the increased demand.

FOR THE COUNTERTOP PROFESSIONAL, ready-made foam sink knockouts can be used to quickly assemble a mold.

TABLE-MOUNT vibrators clamped to pouring tables insure even vibration.

THE FLOOR OF THE Cheng Geocrete® shop.

Sealers. Fabricators are still searching for the Holy Grail of sealers: a sealer that protects as well as a topical sealer, but, like a penetrating sealer, doesn't obscure concrete's natural beauty under a coat of resin. Thus the debate continues concerning the merits of topical versus penetrating sealers. Penetrating sealers don't protect concrete from acid stains as well as topical sealers do, while topical sealers scratch, can burn, and put a coat of plastic between you and the concrete.

We use a sealer that is a hybrid: It penetrates, and though some remains on the surface, it's not enough to obscure concrete's natural feel and look. It gives more protection than a penetrating sealer, but like a penetrating sealer and unlike a topical sealer, it can be reapplied to itself without the need to strip the old layer first.

Considering a Concrete Sink

Be aware that a concrete sink may not wear as well as sinks made from harder materials such as porcelain or stainless steel. Also, some building inspectors are now requiring an abrasion test—which can be both time-consuming and expensive—of a concrete sample before they'll permit a concrete sink. Before you plan and pour an integral sink, check the local building code. A bathroom is a better setting than a kitchen for a concrete sink; in a kitchen, the assaults of soap, cast-iron pots, and scrubbing can rapidly take a toll. We placed close-knit stainless-steel mesh just under the surface of the concrete in the Natoma Street penthouse sinks, as added reinforcement to slow water erosion. Inlaid stone or tile would serve the same purpose.

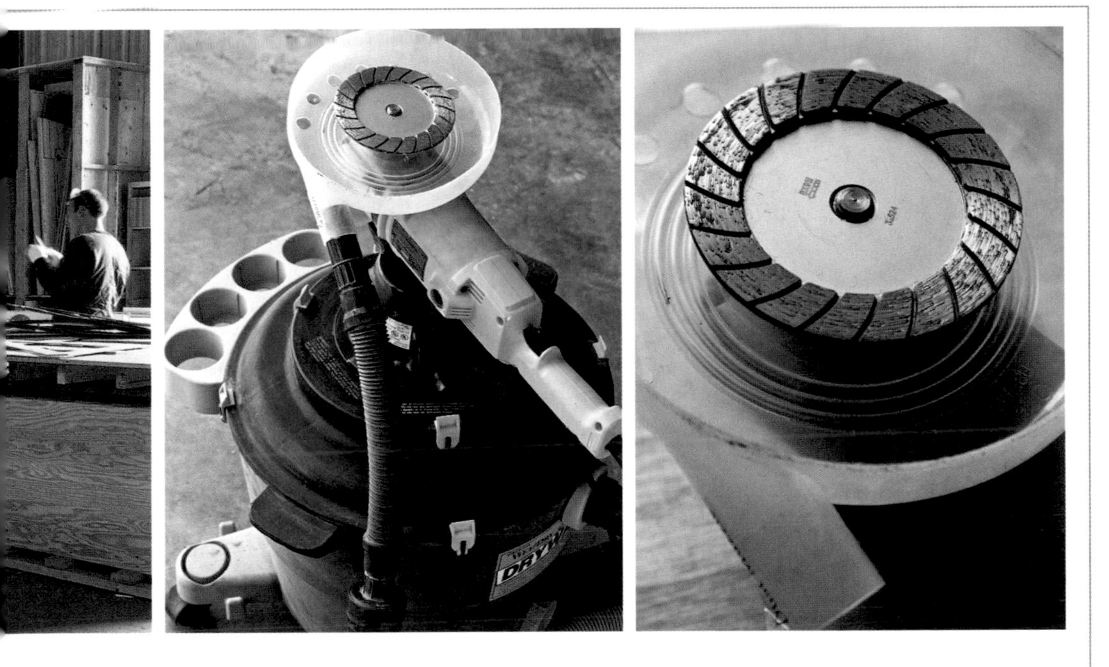

THE DUST-MUZZLE DRY GRINDING exhaust shield directs dust to the shop vac.

DIAMONDS IN METAL GRINDING WHEELS can aggressively remove concrete from surfaces.

Grinding and Polishing. New handheld grinder/polishers, ideal for a small project like a countertop, have come onto the market, so you have a number of good products to choose from (see Resources, p. 208). Steel-mounted diamond wheel pads, which have been around a long time for specialized stone work, are finding their way into our toolboxes. They're especially useful for touchups on walls and countertops poured in place as a single unit (see the Meteor Vineyards project, Chapter 7). The steel-mounted pads let us aggressively grind out trowel marks and quickly create an even, flat surface ready for polishing with the standard resin-mounted diamond pads.

TRADITIONAL IN DESIGN does not necessarily mean old. Here, a massive hood in a new kitchen by designer Dave Condon reflects the scale of a kitchen in the style of an English manor.

AN EXISTING FIREPLACE was incorporated into this kitchen renovation. The drop-down front covers old brick (right). The concrete wall is an end-cap to the pastry counter cabinets. The marble rolling slab was inlaid into the concrete counter.

Reinforcing Materials. One new reinforcing product is a mesh made of carbon fibers (see Resources, p. 208). This mesh lies very flat and is quite thin and easy to use. It works well on thin countertops and in tight situations where rebar is not feasible. We've begun using it routinely in conjunction with rebar, and occasionally without rebar in pieces that don't have overhangs. Better yet, the carbon won't corrode if water should penetrate into concrete. About $1.50 a linear foot in 3-ft.-wide rolls, it's more expensive than rebar or remesh, but well worth it for some applications.

The Tea Shop

Designing in concrete and exploring new forms is always invigorating. I designed a tea bar at Téance/Celadon in Albany, Calif., to be a single sculptural shape that anchors the entire store. It's an example of the way that the design and fabrication of finished concrete can be pushed by form

THE TEA ROOM IS DESIGNED to invite people to sit, relax, and enjoy themselves. Concrete throughout the space connects each element: the floor, the counters, the tea bar, and a water piece.

«**THIS TEA-TASTING BAR** was designed to represent the shape of a traditional Chinese tea cup and was cast in a plaster mold on site.

TO MAKE THE TEA CUP PROFILE for the form work, an adjustable revolving screed was devised by craftsman Hans Rau to smooth plaster over a wood and wire armature.

THE CENTER KNOCK-OUT WAS MADE from a large Sonotube® (cardboard) ring and a foam void piece designed to create an internal shelving space for the tea bar.

THE PLYWOOD TRIANGLES clamped to the formwork "dam" the concrete to form a sloped surface for the drain area.

and function. This flared cylindrical form was inspired by a traditional Chinese teacup. It integrates wall and countertop and was cast on site in a single monolithic pour. Flaring at the top, it serves as both a serving counter and a drain-board for the elaborate tea service, where customers can sit and sample teas the tea master prepares. Because the end result is so massive and unexpected, this fusion of wall and countertop has undeniable visual impact.

The Natoma Bathroom

A countertop with an integral sink is the ultimate expression of sculpture and function. But concrete sinks are not something we do often because concrete is not the most durable material for a sink. Water and abrasives will wear through the finish surface to eventually expose gritty sand. In general, bathroom sinks endure far less abuse than kitchen sinks, so consider a bathroom sink if the urge for

THE WOODEN FRAMEWORK looks like the ribs of an airplane wing. The sink's shape was formed with ¾-in. plywood.

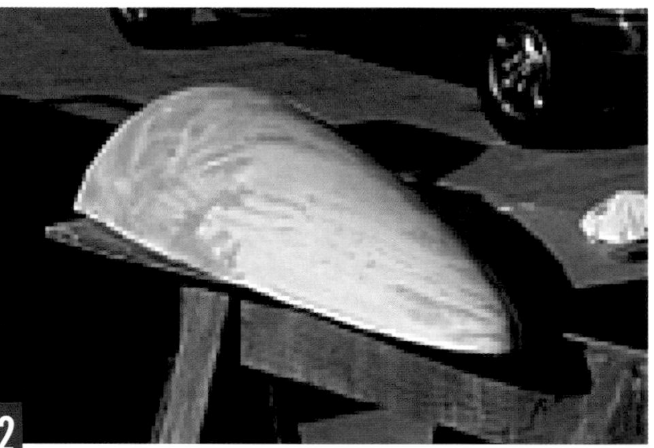

WE LAID RIGID FOAM between the plywood ribs, then added a layer of plaster and sanded the entire form for a smooth surface.

AS A LAST STEP, we painted the sink form with lacquer and put it into the mold.

HERE, A VOID MADE with the rigid foam was fit into the form. The void fills some space under the countertop to lighten the piece.

THE NATOMA STREET MASTER BATH vanity sink spans nearly 12 ft. Three sections are held between walls by a single I-beam behind the dropped-front edge of the countertop.

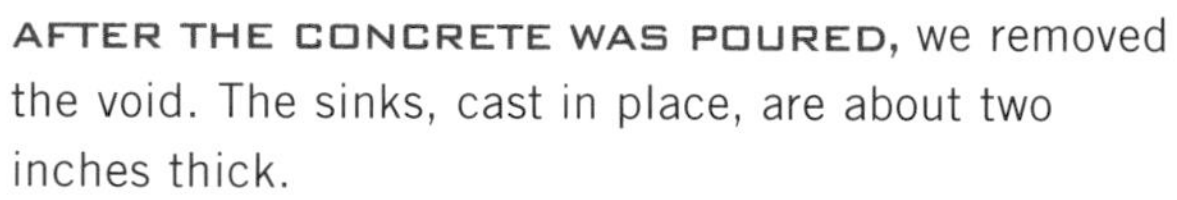

AFTER THE CONCRETE WAS POURED, we removed the void. The sinks, cast in place, are about two inches thick.

NOTE THE FRONT OVERHANG has a channel to receive the supporting steel.

THE MIDDLE SECTION cantilevers over the drain holes of each sink.

THE SECTIONS KEY INTO each other and are sealed together with silicone.

FROM THIS VIEW of the Natoma Street vanity, it's easy to observe the gentle shape of the two sinks as they combine to soften the angles created by the countertop and other square forms in the room.

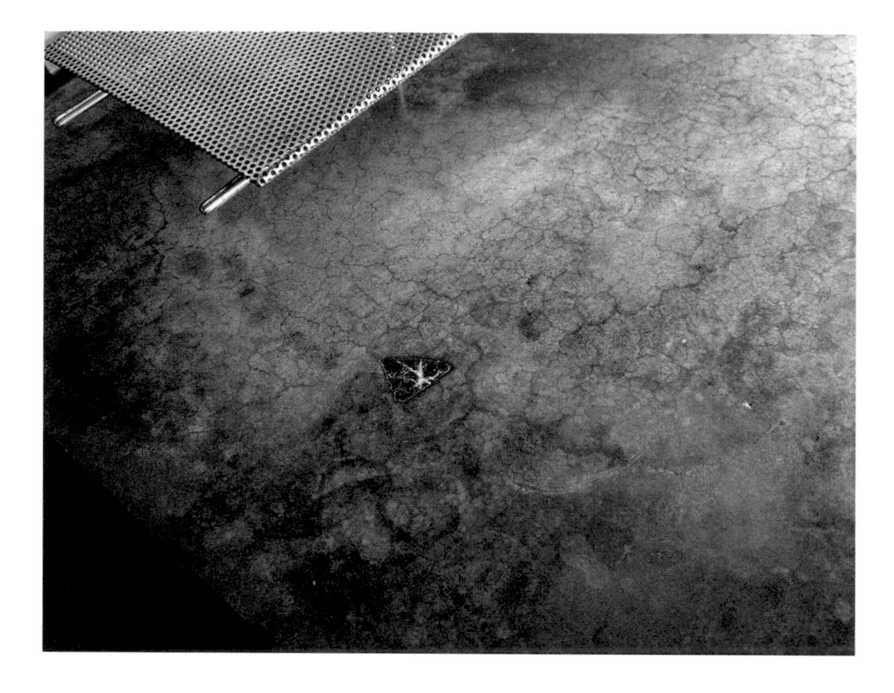

CONCRETE IMPROVES WITH AGE, as this countertop from the McMillan project shows eight years later. The craquelure—the network of tiny cracks in the surface—gives the countertop a translucent depth.

such a sculptural piece is irresistible. We produced such a vanity/sculpture for the master bath at the Natoma Street penthouse in San Francisco (see Chapter 2). The vanity is a fusion of two integral concrete sinks separated by a center platform, or mesa. The soaring vaulted ceiling of the penthouse and the scale of the room required a bold and dramatic shape that would work in the context of the home's interior design.

The countertop is 12 ft. long by 2 ft. wide. It was poured off-site in three sections in a form lined with plastic laminate and reinforced conventionally with rebar and remesh. The mix was our basic stone-colored NeoMix Pro-Formula.

The two faucets drain toward the center drains on either side of a center platform. The entire piece is supported on a steel I-beam hidden behind the drop-down front edge.

Improving with Age

Concrete is a fascinating material because it is so long-lasting, yet seemingly so fragile. Though it is frequently chosen for its strength, its beauty is often its most enduring quality. Crazing often develops on the aging surface of our countertops, and colors mellow into a translucent glaze.

As an artist, I strive, in part, to create design that endures. I find concrete so fulfilling (and thrilling) because it is a product of the design whims of a frozen moment. And yet it contains within it the ability to communicate time passing. The crazing is a reminder of change; the hardness a reminder of its infinite crystalline bonds. Crafted properly, these bonds will outlive us all.

Fixing Small Stains

Concrete is vulnerable to stains from such things as lemon juice, wine, or vinegar no matter what sort of sealer or wax you apply to it. These and other acidic liquids can eat into the fines on the surface and roughen it slightly. Of course, one person's annoying stains are, to another, a lovely patina that reveals years of use.

It's possible to remove individual stains without the need to refinish the entire piece. Below is a step-by-step guide. We recommend passing it on to clients, so they can handle those pesky little stains and abrasions: A stain like may look permanent, but it's not.

THE GEAR YOU'LL NEED: 800-, 1200-, 1500-, and perhaps 3000-grit diamond pads, sealer, and wax.

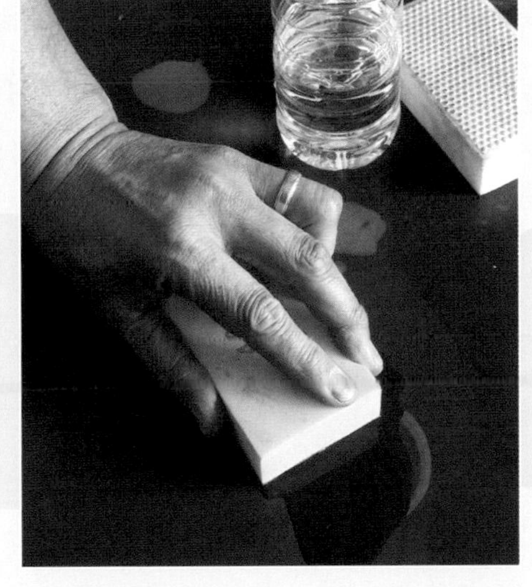

WET THE SURFACE OF THE CONCRETE and start with the 1200- or 1500-grit pad. (There's no need to remove any wax, as the diamond pad will cut through it.)

POLISH WITH A 1200-grit pad until the water gets cloudy. Sponge it off to see how the stain looks. If stain is removed, dry, reseal, and wax.

IF THE STAIN IS STILL THERE, try the 800-grit pad. Work aggressively on the stain, and gradually work out in larger circles (far left). Wipe it clean and check often. There's a danger that grinding too aggressively with the 800-grit pad could expose some of the fines, which will change the look of the surface slightly.

AFTER THE 800-GRIT PAD, left, go over the stain again with the 1200-, 1500-, or 3000-grit pad (the latter will give the surface a shine). Let the surface dry, apply a penetrating sealer, then wax.

PART TWO

THE FUNDAMENTALS OF WALL DESIGN:

5 Character, Form & Expression

Today, our walls are generally framed up with wood, clad with plywood, and then finished with drywall and exterior siding. We build in layers, leaving a hollow core where the utilities conveniently run, and our surfaces are flat, evoking no emotion at all.

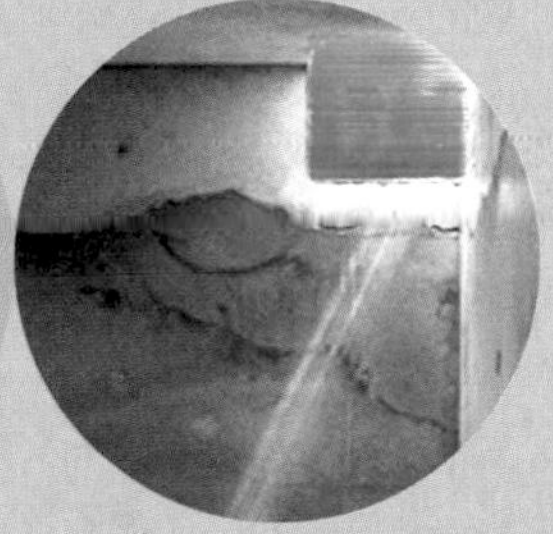

But we still equate "solid" with "quality." To fulfill that desire, a host of products have been put on the market now that attempt to mimic the substance and mass of traditional wall materials—faux plasters, faux "cultured" stone, and thin brick veneers, for example. The manufacturers of these products know we haven't abandoned our primal urge for the solid protection of the wall.

A cast concrete wall, on the other hand, crafted with simple techniques and simple materials, perhaps embellished with a well-considered design element or two—an inlay, some color, texture, or gentle curve—can evoke the emotional bond we have with the massive, permanent, but richly textured rock wall. Concrete has, if you'll pardon the pun, mass appeal.

All you have to do is provide the form. In this chapter, we'll look at wall design and how thoughtful concepts can work in tandem with well-crafted rooms to produce beautiful

« **THIS GENTLY CURVING WALL** acts as both an invitation to the main entry to the house and a channel for rainwater. Rain drains off the roof, then runs down the chains to the caged rocks, where its progress is slowed.

THIS 80-YEAR-OLD WALL in Hong Kong exhibits a refined rhythm of stone and concrete.

IN ENGLAND, THIS WELL-COMPOSED stone wall conveys a timeless sense of substance, craft, and quality.

wall pieces. First, though, it's important to grasp the basic design vocabulary so there's no mystery about what is meant by position, size, shape, the indoor-outdoor context, color, and texture as each pertains to walls. In Chapters 6 and 7, we'll look at how to put this vocabulary into practice.

Origins

At the turn of the century, concrete made with Portland cement was the new miracle material for constructing walls. Builders and architects who'd been using stone saw the potential of concrete right away: It was cheaper and much easier than stone to shape and mold. Bernard Maybeck began to use concrete extensively in his designs. But it wasn't until the Berkeley Hills fire of 1923

BY USING MORE FREQUENT line breaks on the interior "bubblestone" walls than he did on the exterior, Bernard Maybeck created a sense of intimacy for this cozy room in the Wallen House, in Berkeley.

destroyed hundreds of wood-framed homes and other buildings (among them 13 that Maybeck had designed and built) that he started using concrete as a complete structural material, rather than for discreet architectural accents. Maybeck went a step further and developed a material specifically for use in fire-prone areas. He called it "bubblestone," an air-entrained concrete that wouldn't fracture when exposed to temperature extremes.

Maybeck brought a modern sensibility to his designs. In particular, his restrained and elegant use of concrete as an expressive material has been an inspiration for our own work at Cheng Design.

THIS LAVA ROCK IN HAWAII is precisely fitted, revealing a compositional balance of movement and stillness.

IN AN EARLY EXAMPLE by renowned California architect Bernard Maybeck, the walls of the Wallen House are constructed of his own version of air-entrained fireproof "bubblestone" concrete.

⩓ **THIS PENTHOUSE** in San Francisco was designed as a contemporary interpretation of Maybeck's work. The steel-framed windows against the concrete walls are a hallmark of his style.

«**THE PLAY OF** complementary materials—glass, steel, and concrete—in the interior is an echo of the exterior façade.

A Design Vocabulary

A wall offers more opportunities for design interventions than a floor or, for that matter, a countertop. Floors and countertops, after all, have to be flat and reasonably smooth for use, which places a few restrictions on one's design options. A wall, by contrast, can be heavily textured, canted, curved, penetrated, or laced with deep inlays or even protrusions.

CONCRETE AND COMPLEMENTARY MATERIALS

Concrete is a material that anchors and acts as a ballast for other, less substantial materials. Because concrete has its own inherent structural integrity, it works especially well with other materials that have a similar integrity, such as structural wood and glass windows with steel frames.

Deciding how to use concrete with other materials is a balancing act; you don't want to throw too many elements together, and you don't want to strip the setting to the bare minimum. When designing concrete in a home, especially an existing home, don't create an anomaly by inserting something like a concrete wall into a room without real purpose; a massive wall plopped into a cozy bungalow is going to look odd.

Instead, think in terms of connection and function. You might install concrete on the floor of a foyer to create a practical and aesthetic transition from the outside. Then, as a way to continue the visual effect, design a concrete wall to merge with the floor, integrating the materials with the rest of the interior setting.

BY THE WAY

Be aware that concrete is not a good insulator. The thermodynamics of concrete become a critical issue in certain climates. Is the weather very cold in the winter? Hot in the summer? Will the wall receive intense direct sunlight? A thermal barrier of foam insulation can be engineered into a wall to cut down on excessive heat loss or gain.

POURED ON SITE, the island serves as a partition between the kitchen and dining area. A cantilevered cutting board and open shelf complete the clean design.

SIZE AND SHAPE

While concrete can give weight to a room, or even serve as the main feature for a space, it can also play the role of a supporting actor. In the Hertz residence in Berkeley, Calif. (photo on p. 107), a small, pre-cast wall was poured off-site and installed in the kitchen. This wall serves as a visual and tactile anchor for the cabinets, completing the triad of floor, wall, and cabinetry. This is an example of a modest, completely manageable pour that makes a big impact.

It's not at all difficult to take the simple wall and put a curve or slight arc into it (see Chapter 6 for the form detail). In some instances, a curved wall can fit the ergonomics of a space more effectively than a straight wall. In a kitchen in Los Altos Hills, Calif. (left and facing page), for example, which is a tight space, the circulation pattern defined the shape of

SMOOTH CONCRETE AND LIGHT WOOD work in tandem to create a visual play of joinery.

IN THIS LOS ALTOS HILLS KITCHEN, an elliptical island eases the tight passageways while functioning as a space for food preparation. I-beams support the glass shelf on one side and the cutting board on the other, and connect to the post.

BY THE WAY

Some colors don't hold up well outdoors. A sealer may slow the fading process. Also, consider a northern orientation, if possible, so the wall receives less sun exposure.

the wall: a curve that eases the narrow passage between the wall and the island.

In kitchens, the minimum width for a work area passage is about 36 in.; the ideal is 42 in., and there should be no more than 48 in. between adjacent work surfaces. A curved wall, supporting a curved work surface, allows you to push these minimums and maximums in some areas. At the apex, for example, you can allow for less than 36 in., while a straight wall in the same setting wouldn't work nearly as well. A curve is especially appropriate where you have lots of people moving about in close quarters; it's a "friendly" shape that's free of corners to bump into. Another bonus is that

A MASSIVE INTERIOR WALL CAN OFTEN BE SOFTENED WITH DETAIL. Inspired by the Ch'an painting masterpiece, *Five Persimmons*, this wall frames a collection of ocean smoothed stones.

IN THIS BERKELEY HOME, the dark concrete floor rises up to fuse with the light pantry cabinet, a visual anchor for the cabinet, wall, and floor.

Innovations in Concrete Block

Concrete block has always been popular for walls because it is easy to form and work with; there's no need to worry about blowouts as no form is required. Clever design transforms this ubiquitous material into the building blocks of a beautiful home below by architect Wendell Burnette.

The house is located in the high Sonora Desert outside Scottsdale, Ariz., a setting where monolithic concrete walls, without additional insulation, are possible. The shape of the walls was produced by staggering each course of blocks, and by adjusting the curves block by block.

A slurry coat of sand and cement was brushed over the blocks to disguise the vertical joints and emphasize the strong, staggered horizontal lines.

USING INNOVATIVE TECHNIQUES in concrete block, the house emerges as a sculptural form, like the ship-lapped prow of a Viking ship.

CONCRETE BLOCK is placed in ever-widening runs before the concrete is poured into cells.

THE CANTED BLOCK CONSTRUCTION makes curves form from straight units of concrete block.

THE SLOTS, OPENINGS, VOIDS, AND WINDOWS of this wall form a visual composition akin to musical "notes" and become a study in negative and positive balance.

For a ranch-style home in Orinda, Calif. (at left and on opposite page), we crafted an extension of the kitchen to create a breakfast nook. The massive concrete wall grows up from the earth and supports steel-framed windows that by contrast appear quite delicate. Inside, benches are cantilevered off the concrete. From inside, one gets the feeling of sitting on a rock promontory, while the concrete wall creates a nestled, secure feeling. The nook serves as the focal point for family dining; the same approach can work in a variety of settings, such as the kitchen, bedroom, or outdoor rooms.

COLOR

The pigment needed to integrally color the sometimes large volume of concrete in a floor or a wall

THE OUTSIDE OF THE ORINDA EATING NOOK addition sits on deep concrete footings. A whimsically-placed planter still contains the moss that was planted there some 14 years before.

THE INTERIOR WALL of this San Francisco penthouse kitchen supports the windows and visually anchors the interior.

the gentle curve of the wall softens the boxy effect of cabinetry. Also, a wall with an unexpected shape can act as a focal point, drawing the eye from, say, less aesthetically-pleasing appliances.

CONCRETE INSIDE AND OUT

Concrete doesn't rot, it's virtually impervious to water, and it doesn't burn easily, making it a perfect material for use outdoors. And given its potential for warmth and expressiveness, it's ideal indoors, too. In fact, walls that engage both the interior and the exterior present an exciting design opportunity. When done properly, the result is an emotional one: There is an easing of that transition between indoors and out.

GLASS AND CONCRETE COMBINE to bring stunning views into the kitchen nook of this remodeled '60s ranch house in Orinda, Calif. It is fitted with recycled industrial steel windows, and the orientation of the seating area was carefully considered to avoid midday sun exposure.

^ **THE WALL OF THIS PARK CITY,** Utah, kitchen appears to lock right into the butcher-block countertop and support floating cabinetry. Poured off-site (and carefully shipped), the wall features an inlay of recycled steel plate, hinting at a past life.

can be very costly, depending on the type of pigment used. That said, small accent walls are often colored integrally.

There are many integral pigments on the market now for floors and walls. In general, these are powders or liquids that are added directly into the mixer truck. Various combinations of pigments are added in precise amounts to create specific colors. Solomon Colors and Davis Colors (see Resources, p. 208) are two of the several companies that now make colors and mixing systems.

We commonly use at least some carbon black in all our walls to darken the concrete. When we use other colors, we strive for hues that are earthy and natural, rather than bright, artificial colors. But there are ways to add color to a wall other than integrally, some of which are simple, economical, and provide ample opportunity for creative expression.

Combining color with an acid wash. On the one project in Park City, Utah, a small wall in a kitchen, we added ½ lb. of black per 100 lb. of cement and about 10 lb. of blue to the mix. After the concrete cured, we acid-washed it with an off-the-shelf palm-green solution, creating a subtle blue-green. We applied the acid wash with a plastic garden sprayer and then wiped the wall down. It's difficult to control dripping and streaking when acid washing a vertical surface, but this wall was cast off-site and laid flat for spraying, so streaking and drips weren't a problem.

Layering color. Interesting effects can be achieved with just a little integral pigment. On one interior wall (facing page), we colored a small amount of concrete with an intense concentration

of blue pigment (the blue portions were overloaded with color, about 30 percent, three times the recommended concentration). We then filled the form part way, added the colored concrete, and filled the rest of the form. The finished effect is like a geological striation, suggestive of a colored band in a rock cliff face. On an exterior wall, it's even more dramatic (see p. 114).

Because it was important to keep the layer of colored concrete more or less intact, a relatively stiff mix was used and deliberately under-vibrated. This not only preserved the striation, but resulted

THE ISLAND BASE for this Piedmont, Calif., kitchen features a Cheng style signature: a partially fractured ammonite inlay in a mottled ultramarine blue pigmented wall.

BY THE WAY

It's very difficult—and messy—to try to color a wall with acid washes, but not impossible. Likewise, while grinding and polishing a floor or countertop is a relatively simple matter, it's extremely difficult to do the same to a vertical surface.

A Study in Color and Texture

The original entry to this Mammoth Lakes, Calif., house, nestled in the foothills of the mountains, was often buried under 12-ft. snow drifts in the winter, so we designed a large second-story addition to hover over a new, all-concrete "decompression" foyer. With soaring peaks as a backdrop, a fissure of intense blue pigment suggests a striation of mineral deposits. Stacked plate glass, which acts as a skylight and a lens-like window, echoes the thickness and density of the wall that surrounds it. The stained cedar siding applied to sections of the house complements the textured plum-red concrete walls. Our custom-made form ties produced a composition of holes through the wall, adding detail to the texture.

THE GLASS WINDOW was created by fusing individual plates with transparent epoxy.

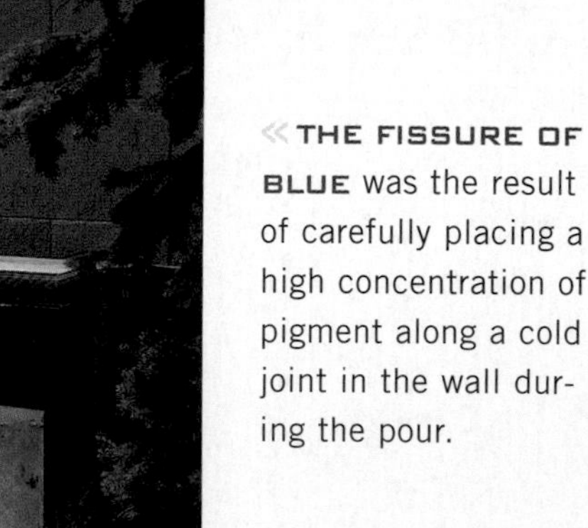

THE FISSURE OF BLUE was the result of carefully placing a high concentration of pigment along a cold joint in the wall during the pour.

THE WALLS ON THE BACK SIDE of the house are of a plum-red mix and molded against a form lined with a heavy-duty plastic tarp, which trapped air between itself and the form, creating the occasional wrinkle in the concrete.

» **IN A JUXTAPOSITION** of textures, a wood-grain pattern adorns the concrete wall surface while the floor is smooth and polished.

in honeycombing, which added to the textured, organic effect we were trying to achieve. The voids also reveal the mass of the concrete.

Combining color and texture. These elements work together in many interesting ways. To color a heavily textured exterior retaining wall, for example, we applied bands of red by brushing water-soluble white glue onto some of the form boards, which had been offset (see below) to create the extreme relief. While the glue was still tacky on the boards, we sprinkled them with red oxide pigment. The glue dissolved in contact with the wet concrete, transferring the pigment to the wall. This color transfer method gives color a faded, slightly imperfect, and mellow look.

^ **THE GARDEN WALL** and fire pit of this contemporary home feature board-formed bands of faded color to effect an aged look.

TEXTURES

As with a form for a countertop, the materials you use for the form or the mold for a wall will directly impact the wall's finish, and thus its function and feeling. Concrete will take on the texture of whatever it's poured against with surprising resolution. Your choice of form materials provides opportunities for plenty of creativity.

Before the introduction of plywood, most concrete was poured into forms made with board lumber. Now we have sheet materials that allow us to create finishes in concrete that are smooth, hard, and glasslike. These smooth finishes work against expectation, since most concrete walls are rough.

Also, in some settings, rough concrete surfaces are simply not appropriate. In a kitchen or dining

CREATING AN IDEAL indoor/outdoor setting, this wall's rustic look was toned down when we added concrete so less rock showed.

TOO MUCH OF A GOOD THING: The original stone and concrete wall lacked rhythm or even a certain artful disorder.

room, for example, where there's lots of activity, especially where food is being prepared and served, easily cleaned smooth surfaces are much better than textured ones. On the other hand, a setting with a lot of glass-smooth concrete can produce a cold, rather forbidding feeling, so it should be used judiciously.

Lining the form. We often line our forms with a plastic laminate, which produces the smooth finish we like. The pressure of concrete in a vertical wall form presses the material firmly against the sides while it cures, and the result can be a very slick surface, when all goes well. Plastic laminates and similar materials also release from the concrete easily, eliminating the need for form-release oil, which can stain or discolor the concrete.

Melamine also makes good form material. Like plastic laminate, it releases easily, and is considerably cheaper. But it doesn't produce a finish that is as glass-like as that produced by laminate. Because we usually grind and polish our countertops, the slight texture left by melamine is not a drawback. (Concrete poured on slick surfaces in countertop molds, unlike walls—which generate high pressure

UNLIKE THE STURDY NATURE of the entry wall that opens this chapter, this wall was designed to erode as rainwater from the gutter above flowed onto it. Mixed with a weak recipe of sand, aggregate, and cement, the eroding section revealed hidden treasures after more than a decade.

IN 1994, before any erosion had taken place, the wall presented a monolithic welcome to this Sebastopol, Calif., home. Only the author knew what was embedded: bowling trophies (see p. 23).

THE WALLS OF FRANK LLOYD WRIGHT'S TALIESIN WEST are highly textured works of sculpture, created when rock was loaded into the form and concrete poured to fill the voids.

—yields very inconsistent, sometimes blotchy surfaces, which then need to be ground and polished.) Because a wall can't be ground or polished very easily, plastic laminate may be the best choice if you want a hard, slick surface.

But experiment. Try forms made of wood or other materials. Form liners can be made from just about anything that will release when the forms are pulled away.

On the McMillan house in Danville, Calif., we used roughsawn 1-in. by 6-in. boards to make the form. The wall is a retaining wall about 5 ft. high, 20 ft. long, and 6 in. thick set on an L-shaped footing, with a fire pit that also serves as a bench. To exaggerate the texture, we drove shims behind some of the boards to push them away from the bracing to varying depths and fitted the boards loosely, which allowed concrete to seep between them, adding still more relief.

Texture outside of the box. You don't need a form liner to create a textured look for the wall every time. In fact, when Frank Lloyd Wright built his home Taliesin West (above) in Scottsdale, Ariz., he built walls that seem to defy the forms that held them. With the help of unskilled interns, he stacked indigenous rock into the wall forms and allowed the concrete to flow naturally around the rocks.

By comparison, one project, an A-frame house in Portola Valley, Calif., presented a wall where the irregular rock surface was at odds with the more refined renovation that was in progress (see p. 117). In order to calm the rough texture, sand and cement mortar with a slight amount of black pigment—no aggregates—was used to fill in between the rocks. When the surface was smoothed, the rock's exposure was reduced. In each instance, it's important to consider context; such a rustic wall might have been perfect in one setting, but in this case, given the refinement of the house, it was clearly discordant. Filling in spaces between the stones subtly reduced the scale.

Translucency, the opposite of texture. If texture is about tactile appeal, then translucency is about visual appeal. For the Natoma Street penthouse bathroom in San Francisco, we created an interior that represented a series of refracting surfaces. The wall to the left of the sink is smooth sandblasted panels of glass. The countertop of slumped glass glows from lights embedded in the two supporting C-channels. The translucent sink is formed from resin and contains coral treasure that's revealed only in the light. The biggest accomplishment, however, is the wall. Comprised of tempered glass set into concrete (then shattered in the mold), it sparkles.

WE CAST A POLYURETHANE SINK with coral floating in its depths. Twin steel C-channels support the sink and the monolithic counter of slumped (re-melted) glass.

A STUDY OF LUMINOSITY, nearly every surface reflects or refracts the light in this bath at the Natoma Street penthouse.

BY DESIGN

CONCRETE ACCEPTS a variety of materials as inlays, which can be an expression of art, whimsy, or personal fancy. Inlays serve as accents that show off the monolithic nature of concrete, reinforcing the idea that it's a single, massive material, not several materials layered one on the other. In a bathroom at the Natoma Street penthouse (right), the glass remained intact against the concrete when we pulled the form away; in a sense, the glass is a monumental "inlay."

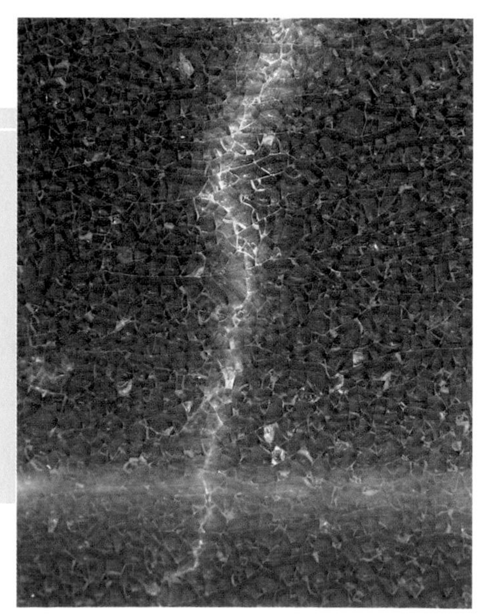

WALL ESSENTIALS:

6 Fittings, Forms, Ties & Techniques

I like to think of a concrete wall as a concrete countertop turned on edge because so many of the techniques we use to design and fabricate a fine countertop are applicable to a fine concrete wall. Like a countertop, a wall is poured into a mold, and whatever shapes, textures, or design elements that are crafted into the mold—inlays, voids, or textured form liners, for example—will be picked up faithfully and preserved in the concrete wall. These are all simple, economical design touches that can result in truly beautiful effects.

With this in mind, you can look at every concrete wall as an opportunity to make an appropriate design statement—simple or complex, cool or flashy—but in effect, we are elevating the craftsmanship and design concept to a higher level, making walls an expression of unexpected artisanship.

But a wall differs from a countertop (and for that matter, a floor) because it requires extra precision in form construction and engineering. In this chapter, we'll focus on the basics of wall design and fabrication. In the next chapter, we'll look at the Meteor Vineyards kitchen in Napa, Calif., a sophisticated project that integrates wall, countertop, and sink in a single pour, a synthesis of elements that creates a single sculptural piece in the house.

« THE AWARD-WINNING HOGAN-MAYO residence in Rancho Santa Fe, Calif., embodies the concept of concrete at home. The cast entry staircase wall leads to an interior of polished concrete floors, cast countertops, and concrete partition walls.

BERNARD MAYBECK CARVED INTIMACY out of an interior of massive elements, blending warm concrete walls, oversized concrete beams, and touches of wood.

Structural Matters

There are several structural matters to be considered as you plan and design a wall, no matter how simple or sophisticated. First, there's the matter of weight. A wall concentrates considerable weight on a small footprint. What will support it? Second, there's the matter of function: Is this a retaining wall that must resist the force of the soils it's holding back? Is it a structural wall that must support the weight of a roof or a floor above? Third, there's the matter of the structural integrity of the formwork itself—anyone who's had a form bulge then blow out under a load of wet concrete knows well the importance of form construction and bracing.

All three of these design and fabrication challenges need to be dealt with; if you have any doubts about your design, the integrity of supporting structures, or about the form, please consult with a concrete professional or an engineer.

WEIGHT

The weight of a wall poured on a slab on grade generally won't be an issue. But weight will be of concern if you're pouring over a wooden subfloor. A wall may not weigh more than a countertop or a floor of the same volume, but the weight of a wall is concentrated over a footprint of only a few square feet. And that concentrated load could cause deflection in the floor supporting it, leading to excessive cracking in both the wall and floor. Another consideration will be the orientation of the wall to the floor joists: The weight of a wall that runs perpendicular to the joists will be distributed more effectively across the structure than the weight of the same wall running parallel to the joists, where the weight is concentrated along a single joist, or on the floorboards between the joists. In other words, carefully investigate the site conditions before designing a wall, and if you have any doubts about the subfloor, get an engineering report.

LOADING AND AREA MEASURE

The same weight of concrete exerts a different load weight when the footprint or area changes. Walls concentrate loads onto a smaller area, so they impose greater loads.

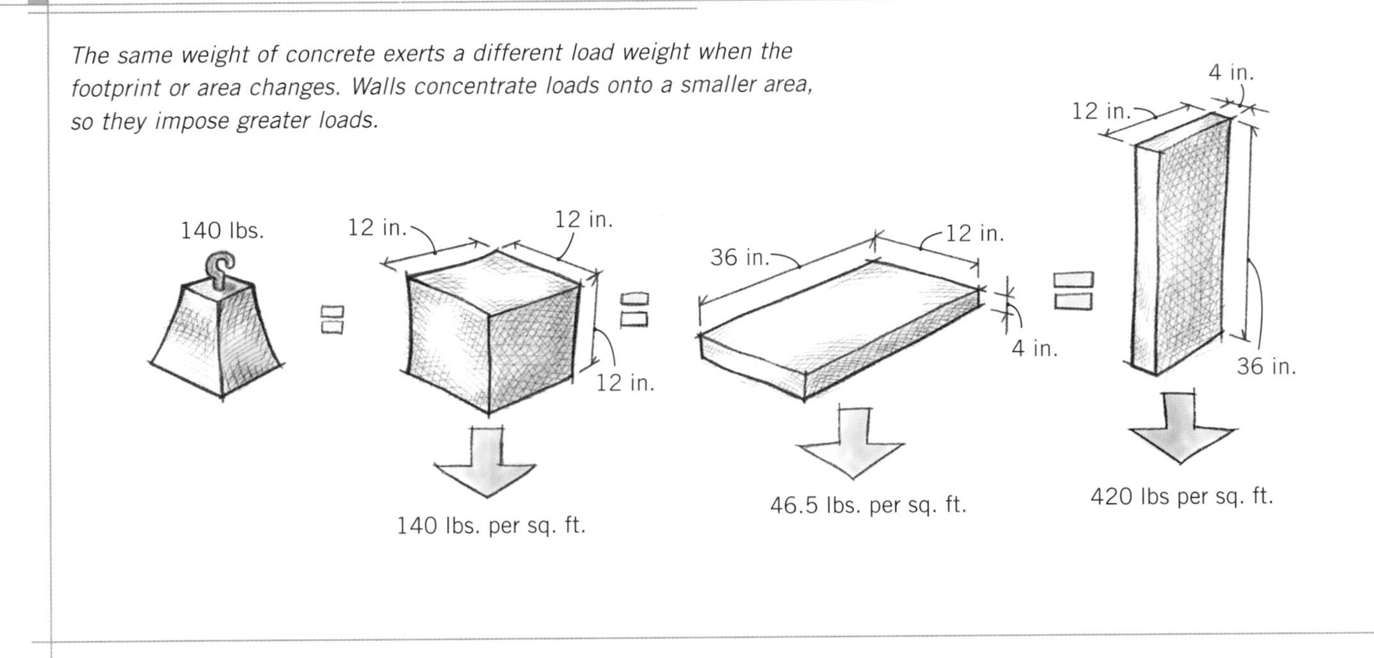

In some cases, where feasible, we've poured engineered footings below the subfloor for an interior concrete wall. For a long wall, we may pour a series of posts along the length of the wall. The frequency of these supports and their depth will depend on the height, thickness, and purpose of the wall. (Is it a bearing wall? How is the weight distributed? Does it carry a countertop?) These footings, as we've fabricated them, typically penetrate the subfloor and tie directly into the wall, effectively separating the wall from the subfloor so that the wall's weight is not a factor

Another way to handle the weight factor is to lighten the load with lightweight aggregates, or with thin hollow walls, or a combination of both. Lightweight aggregates can reduce the weight of concrete by as much as 30 lb. per cubic foot, a 20 percent reduction.

A hollow wall or a wall formed around a core of foam can significantly reduce the weight. This is a suitable option for half walls, partitions, and most non-structural situations, but it may not be an option for a bearing wall.

REBAR AND REMESH are essential materials for reinforcing floors, walls, and countertops.

LIKE THE PROJECT in Chapter 5, this kitchen work station wall contains intensely pigmented concrete slurry that was blended randomly in the mold during the pour.

BY THE WAY

Integral pigments are measured as a percentage of the total weight of the cement. Because pigments tend to weaken concrete, their total weight should not be more than 10 percent to 15 percent of the weight of the cement powder in a concrete.

FUNCTION

In some cases, local building codes may stipulate input from a structural or civil engineer. Most cities require engineering for a retaining wall that's more than 6 ft. tall, since the underlying support systems must be designed and built to withstand the lateral forces of, say, several tons of dirt. Likewise, codes may require an engineer's involvement if a wall will support a vertical load from a roof or a floor above.

The engineering report will specify such things as the amount and placement of rebar, size and

When Concrete is Light

A hollow wall with a foam core is the sort of thing you can do to form a relatively lightweight *non-structural* wall. For a fairly thick wall, we wrap rigid foam around a hollow wood frame. For a thinner, two-sided finished wall, we might simply "impale" the foam on some vertical rebar posts.

You need at least 2 in. of space in which to work the stinger and vibrate the concrete sufficiently, so we attach the rebar or remesh directly against the foam—although conventional placement of the remesh is in the middle of the form.

FORMING THE HOLLOW WALL

A hollow wall still looks and feels substantial, but it weighs less and requires less concrete.

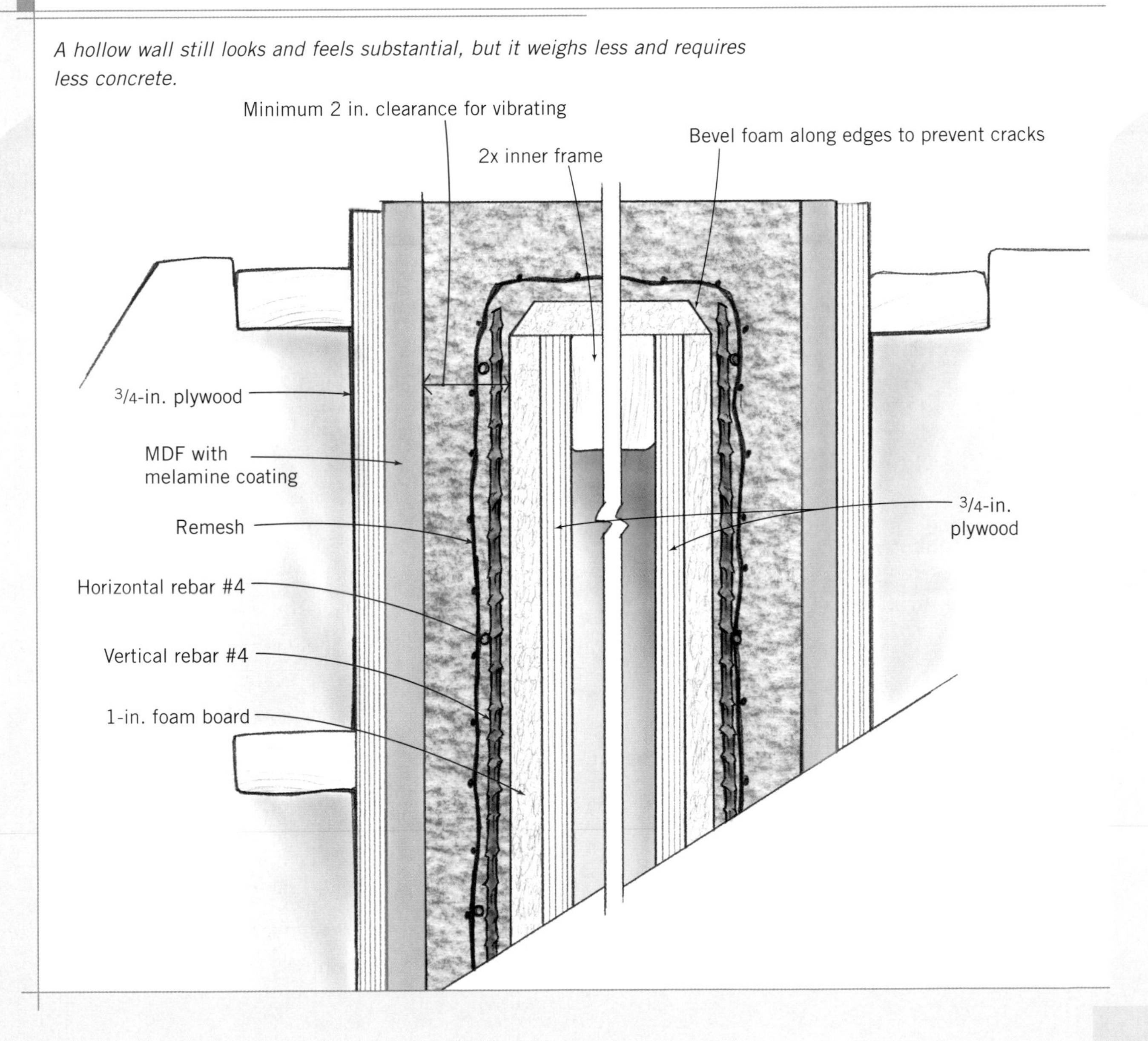

» **THIS FORM TIE** is a basic type for set widths used with 2x material in building foundations.

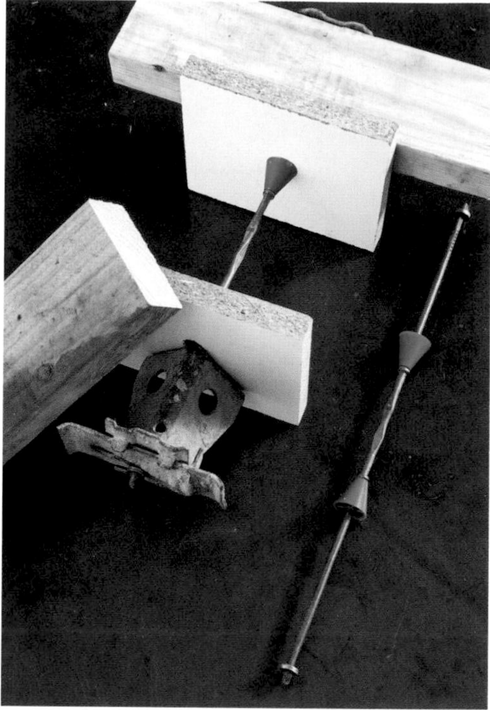

^ **A STANDARD FORM TIE** leaves an inverse cone-shaped hole on the face of a poured wall. They are made to standard wall widths: 4 in., 6 in., 8 in., etc.

spacing of the rebar, and the minimum compressive strength of the mix.

The subject of rebar engineering is huge, certainly too vast to cover here. Concrete has great compressive strength but little tensile strength, meaning it will crack if subjected to bending or twisting forces. The proper size and amount of rebar will increase the durability of the wall. Also, how rebar is fastened, and in particular its placement at corners, will be very important—improperly placed rebar can actually cause cracking rather than prevent it. Likewise, the placement of rebar in the field of the wall or footing—toward the inside or outside, depending on loads involved—makes a big difference in the strength of the wall. We strongly recommend that you have an engineer design the reinforcing.

Making the Form

Unlike the form for a floor or a countertop, the form for a wall must withstand what can be massive forces from wet concrete and gravity. The height of a wall, not the thickness, determines the pressure on the form, and the pressure rises exponentially with height. If you are pouring a wall exceeding about 6 ft. high, consult a concrete professional on the formwork construction.

BRACING THE MOLD

You cannot go overboard when bracing the formwork for a concrete wall. Most who work with concrete have at least one horror story about a form that distorted from the load of wet concrete,

« **A FIBERGLASS SUPER-TIE,** made to adjust to infinite widths like this one, was used on the Hogan-Mayo project.

« **THIS IS AN ADAPTATION** of a standard form tie using threaded rod so that it can be used for any custom width wall.

BY DESIGN

WE SOMETIMES use a form tie of our own design that leaves behind a clean hole completely through the concrete wall, which we leave open or fill with something such as glass marbles.

Our custom ties, which we developed for the Culebra Island project (see p. 140), allow us to form walls of any thickness, unlike most commercial ties, which, with the exception of the fiberglass rods, are designed for specific wall widths. The form tie consists of a ⅝-in. PVC pipe, cut to the width of the form with a kerf cut along the length, and a ½-in. threaded rod. The rod is slipped through the PVC and the form boards and secured with bolts at each end. To remove the form, the nuts are unfastened and the threaded rod removed. The form boards can be pulled off, and the PVC pipe knocked out. It should come out easily because the kerf allows it to collapse. The result is a clean hole through the concrete wall.

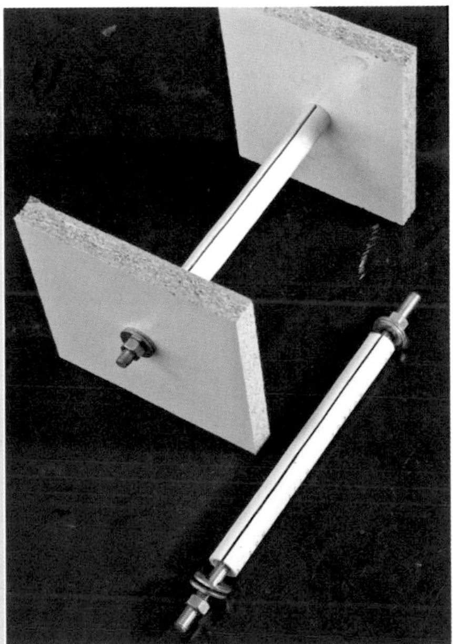

THE HOGAN-MAYO HOUSE, clad in galvanized, corrugated aluminum-coated steel sheeting, looks like a modern-day ark pushing through a sea of tall grasses. The stair resembles a gang-plank, with its canted rail of teak and steel wire.

THE GLOVE IMPRESSION, like a handprint in the sand, is the author's graffiti signature on this pour.

or that collapsed entirely, leaving a grand vision reduced to a mound of worthless mud. We have a few such stories, thus our wall forms tend to be heavily—some would say excessively—braced. There are two basic types of form bracing: internal bracing with form ties, and external bracing with wood bracing and panel gussets. Many projects require both internal and external bracing.

Form ties come in a variety of configurations (strap ties are the most common type, used for rough concrete, such as foundation walls), but they all do the same thing: They tie the two sides together and serve as a tension member between the walls of the form. As the wet, heavy concrete slurry fills the form, pushing the walls outward,

THE HOGAN-MAYO ENTRY WALL was formed with fiberglass super-ties on the horizontal 2x6 bracing: Crew is placing the vertical stiffeners with their cinching clamps.

THE FORM FOR THIS WALL was lined in heavy-duty plastic.

PLAYING ON A THEME of cactus, the fiberglass form ties were left in place and trimmed to protrude about an inch to produce this effect.

WE TRAPPED OBJECTS found on the site, like a remnant of rebar, behind the heavy-duty plastic liner to create ersatz inlays and actual impressions.

IN THE WATTS DINING ROOM, an integral countertop and wall with glass shelf forms a simple L. The conical holes here were formed by a standard tie we customized.

the form ties prevent them from expanding. Sufficient form ties placed at intervals recommended by the manufacturer (with adequate external bracing) are the most efficient and effective way to insure a trouble-free pour. However, keep in mind that most form ties leave behind a hole or, depending on the type of tie used, protruding tabs.

While one doesn't normally want metal or fiberglass tabs sticking out of a finished concrete wall, the conical holes left by some types of form ties, and even the rod ends of some fiberglass ties, can be used as design "punctuation" on finished walls. The form ties must be precisely placed in the formwork in order to achieve the desired effect. Necessity, in this case, is the mother of design invention.

A SIMPLE "L"

The L-shaped wall of uncolored gray concrete (above), part of a kitchen remodel we designed and executed for the Watts home in Portola Valley, Calif., was poured on-site on a polished concrete floor. The wall measures 42 in. high by 60 in. long by 4 in. thick. It was designed to provide a simple and understated separation between the living room and dining room. It also anchors one end of the dining room table.

To pour or not to pour...off-site

We're often faced with the decision of whether to pour on- or off-site. There are many factors that will play a role in the decision. These include the size and weight of the piece: Is it small enough to make off-site and deliver like a sculpture or piece of furniture? If so, then how far away is the site where it will be placed (that is, how costly will shipping be?), and how accessible is the site? Can the piece be maneuvered through doorways or up a tight stairway? If you're considering pouring on-site, then you'll need to consider the situation at the site: Can concrete be delivered to the site, and can the site tolerate the mess of hoses, workmen, wet concrete, formwork?

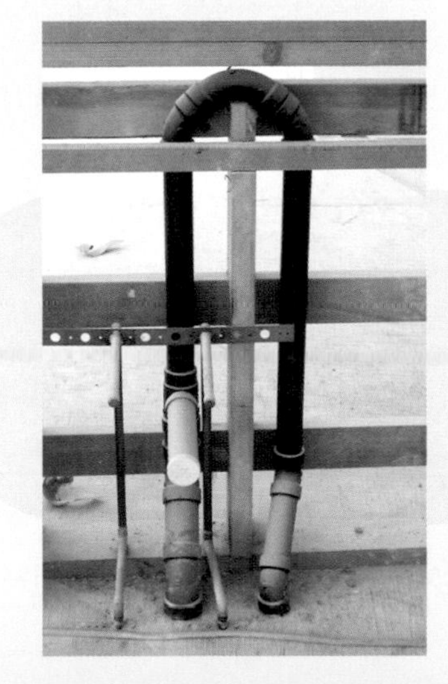

THE WOOD-FRAMED ISLAND WALL with loop vent plumbing is ready to receive an off-site poured wall.

IN BOTH CASES HERE, the advantages of pouring in our shop outweighed the advantages of an on-site pour. The pieces are complex, and in each case, the site could not handle forms being in place for long lengths of time.

THIS ISLAND BASE was mixed with white cement, ultramarine blue, and chromium green pigments. The yellow inlays are hardened concrete remnants that were found lying around our shop and glued to the form.

The Watts wall was constructed with ¾-in. plywood panels, and panels of MDF lined with plastic laminate. This laminate releases easily from the cured concrete and leaves behind a very smooth surface. We formed a knockout for the glass shelf with a strip of rubber, and later inserted into the groove an aluminum channel to hold the glass. Conical form ties were spaced precisely to be both functional and to leave behind holes that would be pleasing design elements (see the Bretton house, pp. 114 and 115, as another example).

Other small touches also contribute to the overall effect. Note, for example, the reveal at the bottom of the wall. This visually separates the wall from the floor, and lightens the look a bit.

BRACING ON A FINISHED SLAB

In the case of the Watts partition wall (see p. 132), we poured the wall in place, creating one particular challenge: How to brace the formwork

FORMING WALLS WITH AND WITHOUT TIES

To form walls without ties, more external bracing is required. Using form ties to hold forms together means that forms can be erected and removed faster and more easily.

Without Form Ties

Concrete
Bar clamp
2x4
2x4
Plywood bracing
2x4 bracing

With Form Ties

Concrete
Form tie rod
2x4
Rod end
2x4
Bracing plywood
2x4 bracing

Note: taller walls will need additional vertical strong-backs.

THE CONCRETE WAINSCOT in the Hogan-Mayo residence is not decorative trim but a 6-in.-thick structural wall poured at the same time as the foundation, with veneer plaster over drywall running the span to the ceiling.

FORMING A WALL WITH AN INTEGRAL COUNTERTOP

This section view shows one way to create a massive-looking concrete wall without using a massive amount of concrete. The structural core is a wall framed with 2x6 studs and anchored to the floor structure with threaded rod. The rod extends through parallel strand lumber (PSL) blocking and through a PSL cleat. Plywood and rigid foam board comprise the form's inner lining. The form for the slanted outer surface of the concrete wall is removed after the concrete has cured.

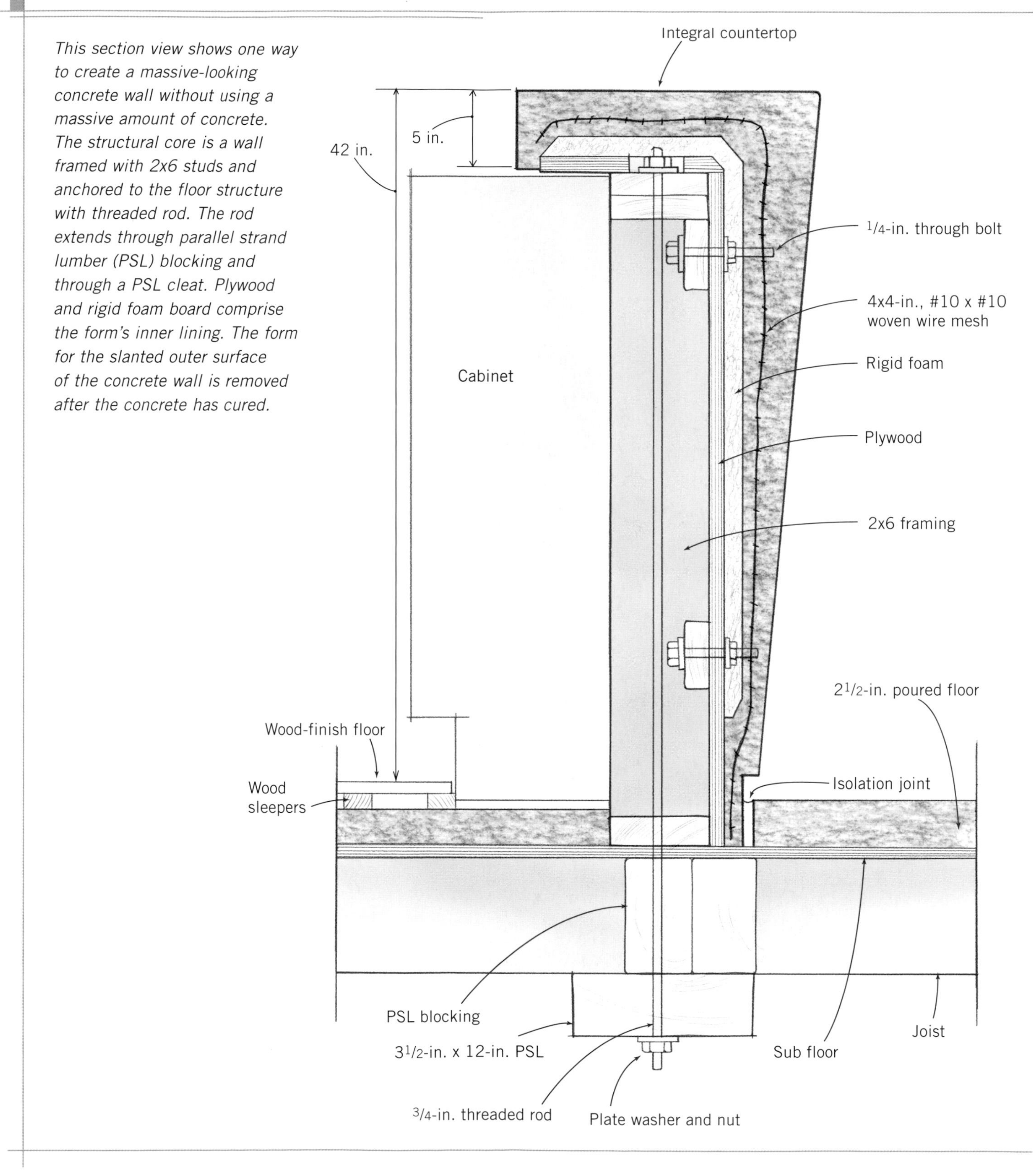

on a finished concrete slab. Even though we planned to use form ties on this piece, the form needed to be braced externally. Normally, we would have screwed each form brace to the sub floor, but here, we couldn't nail or screw all the braces to the finished floor. Working in our favor was the form's stable L shape and the fact that the wall wasn't very high, so we didn't need many braces. To secure the braces we did use without driving lots of fasteners into the floor, we laid down an apron of ¾-in. plywood around the form, using a few concrete screws to secure the panels to the concrete floor. Then we attached the braces to the apron. This provided sufficient stability to the form. When the wall was complete, we removed the bracing and apron and patched the few screw holes with a quick-set mortar.

FORMING A WALL-COUNTERTOP

Monolithic wall-countertops create a dramatic presence in a room. The countertop and wall merge as a single, dynamic structural entity. How to actually support an unstable inverted L and retain the structural feeling is considered case by case. The integral wall and countertop in the

FROM THIS VIEW OF an integral wall and countertop, the craftsmanship is on display: A seam runs where the two pieces are joined.

THE INTEGRAL WALL AND COUNTERTOP was poured off-site in our shop and is supported by plate-steel shelving and wood cabinetry.

THE BLUE FOAM KNOCKOUTS are provisions for electrical boxes and receptacles.

Marion kitchen (p. 137), which we poured off-site, is supported by plate-steel shelving and cabinetry. The steel stabilizes the piece in a visually light way. To support the Meteor Vineyards wall/countertop (see Chapter 7), poured on-site, we cast in small concrete wing walls, which are hidden by the built-in cabinets.

CURVED FORMS

We've fabricated curved walls with a simple, lo-tech method. Materials needed per section are: several sheets of ¼-in. plywood and ⅛-in. Masonite, plenty of 2x4s, fiberglass form ties, and $\frac{1}{16}$-in. plastic laminate.

FROM ABOVE, the complex interlocking of top and wall pieces becomes apparent.

THE TIGHT RADIUS of the form required careful heating of the plastic laminate lining when it was glued into place.

MADE WITH 3⁄32-IN. STEEL, these forms were easier to fabricate than wood forms. However, sheet metal roller equipment is necessary to make the curves.

To make the form:

1. Fix the vertical framework of 2x4s on 6-in. centers (or even closer) to top and bottom pieces of 3⁄4-in. plywood cut to the desired curve.
2. To these verticals, use drywall screws to fasten three layers of the 1⁄4-in. plywood. If the curve is tight, score the plywood with vertical kerfs so the panels bend without cracking. Use Masonite as the final layer to smooth out the plywood.
3. Coat the final layer of Masonite with contact cement. Do the same with the laminate and carefully adhere. The contact cement, plus the pressure of the wet concrete, will be more than sufficient to keep the laminate in place.

STEEL IS AVAILABLE in long lengths, so fewer vertical seams are necessary.

Recently, with Hans Rau, we experimented with curved forms made of 12-gauge, 3/32-in. rolled-steel sheets. The steel form is quite strong, meaning fewer form braces are needed and blowouts are unlikely. And it's possible to create very clean, sinuous curves with steel. The drawback of steel is that you need specialized equipment and some experience to work with the stuff.

The rolled steel surface has a texture of its own, which the concrete will pick up; this may just be what you want, but if not, you can always line the surface of the metal form with a material of your choice, smooth or rough. On the Plummer counter (right), for example, we glued a layer of plastic laminate onto the steel using contact cement.

BY THE WAY

The longer you leave concrete in a smooth mold made with plastic laminate, the harder (and smoother) the finish will be. Normally, for the hardest finish, we leave the forms on for at least 10 days. If you pull off the forms sooner, when water is still evaporating from the surface, the finish will appear dull, though it may still be smooth to the touch. To bring out the sheen, wax and/or polish the surface after it fully cures.

BY DESIGN

FOR OFF-SITE WALL SECTIONS that need to be joined together, we make "dog bone" clamps. One project (see p. 138), involved an 11-ft. curved wall, which we poured in three sections in our shop and then delivered to the site. We cast voids into each section of the wall for the fasteners. Each clamp, about 10 in. long, consists of a length of threaded rod, two washers, two nuts, and two large plywood plates that act as cushions for the concrete.

SLIP-CAST

Because the risk of a form failure increases rapidly as the height of the wall increases, very tall walls, especially in high rises, are often "slip-cast." That is, the walls are poured in sections with moveable forms that are raised up as the previous section of concrete sets up. With a relatively small amount of formwork, large walls can be built.

The Culebra Island Project. We applied this slip-cast idea to the Culebra Island project in Puerto Rico, a house that Cheng Design conceived. The house sits on a promontory overlooking the ocean and, in the distance, the Virgin Islands. To keep the profile of the house low, we set it partially into the hillside, and so we needed to build a long, gently curving retaining wall as part of the house. The wall is 160 ft. long and at its highest is 15 ft.

To start, we plotted the curve of the wall in 8-ft. sections, and produced 20 slip forms, each 8 ft. long and 30 in. high, and consisting of a pair of plywood panels faced with plastic laminate. At the top and bottom of each panel was a key

CAST IN STEEL FORMS, this wall and countertop required fewer braces, meaning less damage to the finished floors.

^ **THE SINUOUS FRONT DECK FORM** awaits a finish topcoat of concrete. The foreground powder room "prow" walls were cast in solid concrete and plastered over with a veneer of concrete stucco.

into which we placed a strip of rigid foam. These replaceable strips produced a shoulder at the top of each pour, into which the keyed strip of foam of the next raised form could fit for support. This foam strip also prevented water and fines from the wet concrete from running down over the previously poured concrete.

Each form was secured with our home-made PVC form ties (see p. 129). We left the forms in place for about a week after each pour, so the curing concrete would remain in contact with the laminate as long as possible.

Since each lift requires days of curing time, this approach takes longer than a single pour, but the slip-cast system saves the costs and risks of an elaborate bracing system. The slip-cast system is a "limited fail-safe system." That is, any form failure won't be a complete disaster, since only one relatively small section of the wall would need to be re-poured.

Before sending the forms to Culebra Island, we poured a sample 16-ft. section near our shop and videotaped the process, start to finish. We then

>> **BEFORE THE WORK BEGAN** in Puerto Rico, two sections of the curved form were tested in lifts.

sent the video to the contractor, Billie Thomas, so that his crew could watch it before assembling these unconventional (to them) forms and pouring the walls. As it turned out, the entire job went off without a hitch.

WALL CUTOUTS, simply voids in the forms, bring in more light and frame the spectacular ocean view. Each horizontal line represents a "lift" of the forms on top of a previously poured course. The rebar in the background awaits the final course.

TILT-UP

Tilt-up walls are prefabricated wall sections that are poured flat on a floor slab or a temporary casting slab, then tilted upright into position and fastened together. These days it seems as if every huge warehouse or strip-mall mega-store is constructed with tilt-up panels. That's because it's a cheap way to build, at least when applied to create basic shells for enormous spaces like warehouses. But pouring a wall flat allows the designer to explore a variety of creative treatments. Essentially, whatever you can do to a floor or countertop, you can do to a tilt-up wall. The downside to tilt-up walls is that you have to figure out a way to raise what could be very heavy panels, which may require special equipment and expertise. For that reason, we've found that the approach makes the most sense for small pieces, such as a counter-height wall, used as an interior accent.

BEYOND THE BASIC WALL:

7 The Meteor Project

In Chapters 5 and 6, we discussed wall design and looked at some essentials of construction. In this chapter, we're going to show how one commission, a kitchen we'll call the Meteor Vineyards project, called on the sum total of our experience. In the end, we included many ideas that had germinated and evolved over the past 20 years, among them, concepts I'd begun to develop working on my own house.

The client trusted me. He said, "I'm giving you the heart of my house. Don't mess it up. Finish in five months." Fortunately, we shared the same vision: It would be a kitchen that was highly detailed, expressed great emotion, and was in complete service to cooking.

The kitchen was the center of a 10,000-sq.-ft. stretched "cabin" with walls constructed of rammed earth, wood framing, and glass. Designed by Cutler Anderson Architects and requiring five years to complete, it called for a concept that matched its scale.

A monolith was the answer—simple low walls and countertops to echo the mass of the house and anchor the space. These elements would need to be tactile and earthy so they would harmonize and contrast with the rammed-earth walls.

« **ANGULAR CONCRETE FORMS** and varied textures define and sculpt the open space of this new kitchen. The concrete walls and counters complement the rammed earth structural walls in this dramatic home.

The cooktop would be a fusion of cast concrete with stainless steel; the sink pre-lined with tile and cast together with the countertops; the concrete counters cantilevered and merged into granite.

In essence, every phase of the building and fabricating with refined concrete came into play. It is presented here in detail so that you can see that it is, however ambitious, simply the result of a step-by-step process that anyone with a little practice and a lot of patience can accomplish. We took the time to do the best we could, and that was satisfying in itself.

Making the Form

Once matters having to do with weight, basic structural integrity, and form engineering have been dealt with, you're ready to build the form.

Constructing the form for a fine finished wall is little different from building the form for any concrete wall. Yet the decisions you make about such things as form materials—slick plastic laminate rather than rough plywood, perhaps—and inlays, can make a huge impact on the final look of the wall, as can the care you take in curing the piece

OUR STARTING POINT: what would become the Meteor Vineyard's kitchen in Napa, Calif.

THE FOAM MODELS were crucial to showing how wall, countertop, and surrounding structures are integrated. Here we see the countertop run and the intersection of the food prep integral sink.

THE COMPLETE MODELS. Formwork and bracing are fitted to the foam model. The wall is canted out 7 degrees and braced externally, without form ties.

COPY-ROUTED PLYWOOD STIFFENERS and pre-cut 2x4 bracing were used to keep the form from bulging or deflecting.

and removing the form. Lots of creativity and some attention to detail can transform the mundane into the sublime.

It can be difficult to visualize how to build a form, especially if it's very complex, with many inlays, voids, and knockouts, like the form for the Meteor wall and countertop. This project included a pre-cast concrete sink run, a structural post, a cast-in-place food prep sink, and a custom stainless-steel cooktop, which all had to fit perfectly together. As a first step, full-scale three-dimensional models of the wall and countertop were constructed out of rigid foam.

We glued the pieces of foam together. Then we put the models in place on-site. In effect, it was a three-dimensional template that we measured, directly transferring all the dimensions onto the off-site molds as well as the on-site wall forms. And that ensured the accuracy we needed when it was all to be put together.

BY THE WAY

Rammed earth, also known as stabilized earth, is a wall system that dates its origins to hundreds of years ago. Engineering advances in the last 20 years have made "earth walls" an aesthetically pleasing, low-maintenance alternative to wood-frame construction. At the Meteor Vineyard, the earth was packed in layers with a pneumatic hammer into heavily reinforced forms.

Plastic Laminate

The mold for the Meteor walls and countertop come in two parts: ¾-in. plywood backing a ¾-in. medium-density fiberboard (MDF) panel faced with plastic laminate for surfaces that would be exposed, and MDF faced with melamine for surfaces that wouldn't.

BY THE WAY

The rebar should be placed no closer than 1 in. from the face of the wall. Any closer may result in "ghosting" on the surface of the concrete, a slight discoloration directly over the rebar. Because it's difficult to polish a vertical surface, there's little you can do to fix this.

The ¾-in. plywood backing was screwed to the braces through the face of the panels. Then we fastened the plastic-laminate and melamine-faced MDF panels to the plywood from behind. By using the ¾-in. plywood, we could screw into the MDF from the back, using as many screws as needed to pull the MDF into place. When it was time to release the form, we removed the screws and pulled away everything but the MDF until we were ready to free it from the concrete.

Inlays

Placing inlays in a wall mold is not much different from placing inlays in a countertop mold, other than the obvious difference that inlays for a wall

1 **THE BRACED FORM WALLS** are screwed to the subfloor to prevent movement at the bottom.

2 **A STRUCTURAL POST,** around which the countertop and wall were formed, is wrapped with ¼-in. packing foam in order to prevent cracking as the concrete cures.

3 **THE PLASTIC LAMINATE** is mounted on a ¾-in. MDF panel, which is then screwed from the back against a ¾-in. plywood form.

« THE FLURRY OF ACTIVITY, checking and double-checking before the pour.

4

REBAR IS WIRED TO REMESH and then hung from wires screwed to the top of the form. After the pour, the wires are snipped so the surface can be screeded and troweled.

5

BLACK CAULK IS RUN DOWN the seams of the wall form and then a stick fitted with a razor is used to scrape away excess. The caulk seals the mold and creates rounded, easy-release corners.

BY THE WAY

The concrete at the bottom of a plastic laminate wall form tends to conform to the slick surface of the laminate more than it does at the top. We suspect that the pressure of the concrete keeps it in contact with the surface of the laminate more thoroughly than at the top. The slight discoloration that can appear at the top of the wall may fade as the concrete continues to cure; waxing or a sealer can help obscure it; and it can be diminished by lightly polishing with diamond pads.

PYRITE CRYSTALS are set in cast polyurethane surrounds. The assemblies are then fixed inside the form with silicone caulk.

THE FINISHED TILE MOSAIC INLAY brings depth and interest to the face of the wall.

WHEN THE FORM IS REMOVED, the rubber knockouts pull away, leaving the pyrite embedded in the concrete. This approach fuses the pyrite into the concrete, so it has a fitted look.

THE AUTHOR PLACES mosaic tiles onto a strip of rubber for the horizontal inlay on the face of the cooktop wall.

THE MOSAIC PIECES are lined up on the rubber strip and held in place with silicone caulk.

need to be fixed firmly to a vertical surface. On the Meteor project, we placed only a few decorative inlays.

For the horizontal inlay on the face of the cooktop wall, I made a mosaic from bits of broken tile and stone, securing them onto a thick bar of urethane mold material with a thin layer of black silicone caulk. The whole assembly was glued to the side of the form with more silicone caulk. The concrete flowed around and between the tiles, holding them in place. The rubber strip pulled away easily when the form was stripped, leaving behind the mosaic recessed in the wall.

As a variation on an integral sink, we created a long, narrow, tile-lined food prep trough. Note that in the top photo on p. 152, the void is turned on its side, and crackle-glazed tiles, cut in strips, are laid out on the insert for size. Before the pour, the tiles were turned over, face against the rubber, and glued in place with a thin layer of black silicone caulk.

COUNTERTOP LAYERS SERVE DIFFERENT FUNCTIONS, and each function in turn results in a different form: The sink and cutting board counters project at different levels into the tile-lined prep sink. A perforated stainless "colander" slides on the counter's edge.

TRANSLUCENT CAST URETHANE embedded with natural coral and other found objects from the beach softens an otherwise sharp corner.

» **AGAINST A BACKDROP** of Plyboo™, the monolith of concrete walls and counters unifies the composition of materials: rammed earth, stainless steel, zinc, plate steel, granite, and urethane.

^ **THIS VOID,** made of MDF clad with stiff foam, creates a long, shallow food-prep trough in the countertop. Tile glued to the foam became the trough lining when the form was removed.

The tiles were locked into place by the concrete, and the rigid foam broke away easily when we pulled the form apart. The alternative approach would have been to create a void in the concrete and mortar in the tiles later, but the way we did it, the tiling is done, no grouting is necessary, and the tiles are an integral part of the concrete, not something applied later.

We didn't tile the bottom of the void, however, but rather applied the tiles later, conventionally. There were two reasons for this. First—note the holes drilled in the bottom of the void—we needed a way for air to escape during the pour, otherwise there would have been a high risk of bubbles creating holes in the concrete under the trough. Second, since the trough drains to one end, we wanted to be quite sure the slope was correct. The best way to do that, we decided, was to create the slope with a mortar bed afterwards.

« **THIS COMPLEX STRUCTURE** holds the void for the food-prep trough in place without touching the wall sides of the form, allowing the entire top to be troweled.

HERE THE AUTHOR and craftsman Rob Andrews place concrete while the rest of the crew check last-minute details.

THE AUTHOR DELIVERS CONCRETE from a 2-in. hose while Hans Rau, master form builder, holds a board to protect the laminate.

DRILLED HOLES ARE CRUCIAL in allowing air to escape from under knockouts, preventing voids.

THE CLEANUP SINK features a drain board with ⅜-in.-thick brass inlay strips and a steel dish-drying rack integrated into the counter wall.

Mixing and Pouring

Concrete for a wall is similar in many ways to concrete for a slab. The interactions with the ready-mix driver, coordination with the pumper, and considerations about time in transit, weather conditions, and so on—covered in Chapter 2—all apply to walls.

But there are differences. With walls, as opposed to floors, for example, there is relatively little exposed surface to finish by hand—just the top of the wall. The challenge with walls is more insidious: the risk that a screw is lacking or a brace is unsecured, and that the forms might fail under the extreme pressure of the wet concrete. One must always keep one eye open and an ear cocked for signs of danger. Better yet, always take your time—no, extra time—to build wall forms.

PYRITE CRYSTAL BLOCKS sit like gems in the man-made matrix of concrete.

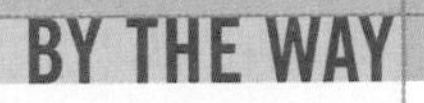

BY THE WAY

Before you begin your pour, have a small screen handy–the fineness of a screen door–to prep some concrete for patching. Push the concrete through the screen and use this cream to fill small voids in the surface and to tighten up around any inlays. It always seems like there is never enough cream at the end to go around.

1 **CAREFUL SCREEDING** of the top of a wall is key to having a level and smooth surface in preparation for troweling.

2 **DECORATIVE AGGREGATES**—turquoise, jadeite, and washed glass—are scattered over a field where inlays, including the author's trademark ammonite, have been pressed into wet concrete.

3 **USING A RICH MIX (6.5 SACKS)** allows you to trowel to a burnished finish. It's important to smooth the concrete as thoroughly as possible, to cut down on grinding later.

MIX DESIGN FOR A WALL

The same basic mix for walls is used for floors and countertops (see Chapter 2), a 6 to 6.5-sack mix, with ⅜-in. pea gravel. Use only pea gravel for walls, since tolerances are usually small. Some of our walls are only 3 in. thick, with rebar and inlays, so the smaller aggregates are needed to create a mix that flows more easily around the various impediments. Other ingredients are the same as those for a floor or countertop: stealth fibers, water reducer, and enough water to create a 6-in. slump. We added a ½ percent load of black to the concrete for the Meteor walls to darken the concrete.

PLACING THE CONCRETE

For walls, a 2-in. hose is best, since it allows you to work the concrete into small spaces. Care should be taken not to let the concrete fall into a tall form, as this can lead to some separation of the aggregates from the fines, with the heavier aggregates concentrated in the bottom. If there is sufficient room to stick your hose into the form, it's always best to do so, pulling the hose up ahead of

BY THE WAY

Pour some concrete into a bucket when it's first delivered. Some of it can be used for small touchups if needed. More important, the concrete in the bucket can be tested later if it turns out there's a problem with the batch. Date and label the bucket.

A CURING COMPOUND slows water loss during curing. We applied three coats here, as the weather was hot, windy, and dry, and the house wide open.

THE POUR COMPLETE, the room is emptied out and cleaned up. The forms will be left undisturbed, for a minimum of 10 days, under plastic tarps held off the surface so that water doesn't condense in the forms.

the rising concrete. Also, be careful not to let the concrete scrape against the sides of the form, especially if you're using a smooth plastic laminate. Any scratches in the laminate will show up on the surface of the finished concrete. Try holding a thin, flat panel of wood or plastic against the side of the form where the concrete is being placed to protect the side.

When pouring a wall in which you've placed inlays, be aware that a thick inlay that extends in from the side can trap air, which will leave voids or blowholes in the concrete. Try to bring up the wet concrete slowly to the bottom of the detail to allow any trapped air to escape out one end. If

>> **THE THEME OF CONNECTIVITY** is evident here where the canted countertop piece, with its 7-in. overhang, is intersected by the post.

1 **AFTER THE CONCRETE SETS UP** and cures, pull away the plywood form supports and the plastic laminate.

2 **SOMETIMES WHITE DISCOLORATIONS** are evident after the laminate is removed. A light polishing will remove the discolorations.

3 **AFTER THE FORMS ARE REMOVED,** the short stud wall remains in place, to be hidden behind the cabinets.

you can, reach into the area and "massage" the concrete to break up any large air pockets. If the void area is at the top of the form (see p. 153), then be sure to drill holes so that the wet concrete itself can seep through, allowing the air to escape ahead of it. Any voids that do form, despite your best efforts, can be filled with an epoxy such as PC7 (gray) or PC11 (white) or a slurry of fines and cement with enough pigment to match the color.

« **FOR A SMALL JOB,** you don't always need a grinder with a built-in water feed. On the Meteor job, to control excess water from damaging the recently poured walls, we used a water-filled container with holes punched in the bottom to keep the piece wet.

HONED, BLACK GRANITE COUNTERTOP and concrete wall against bamboo cabinets make for a pleasant graphic contrast at the small beverage sink. The wall cants out over the floor and engages a pillar at the far end.

Curing

Curing a wall is simpler than curing a slab, since you don't have a large surface area that must be protected from the vagaries of temperature or humidity (for more information, see Chapter 2). Remember:

- The concrete should be kept at 70 degrees to 75 degrees, if possible, for the first three to seven days.
- It should be kept at high humidity to control moisture loss.

For the Meteor project, we used a curing compound on the top of the wall, as the weather was extremely hot, windy, and dry. We applied three coats with a common garden sprayer. We usually don't use such compounds, as they can discolor the concrete. Here, we planned to grind and polish the top of the wall and the countertops, so any discoloration would be removed.

« **CONCRETE IS HEAVY:** It takes most of the crew to maneuver a piece that was cast off-site into place.

» **A CHENG TIDAN VENTILATION HOOD** hovers over four burners, a wok burner, deep fryer, and grill. A zinc backsplash and shelf unit keep cooking surfaces hidden from the dining area.

BY THE WAY

Resist the temptation to touch a new concrete wall before it cures for at least 10 days out of the mold and has been sealed. Often, hands and fingers can leave permanent "ghost marks" on the surface of fresh concrete, especially on a slick finish.

» **THE GEOCRETE® SHOP-POURED SINK COUNTER** with the site-poured wall. The small black squares on each end are electrical outlets set into the concrete face. Spaces between the massive rammed earth walls are a breakfast nook (background, right) and a children's play area (background, left).

PART THREE

HEARTHSCAPES: Mantelpieces & Fireplace Surrounds

8

The attraction of a fireplace is probably ingrained in our very cells; light a fire in one and sooner or later everyone will gather near. Or as noted architect Christopher Alexander put it in his seminal work, *A Pattern Language,* "The need for fire is almost as fundamental as the need for water." Properly contained in a stove or fireplace, fire gives life and light and warmth. It completes our idea of home.

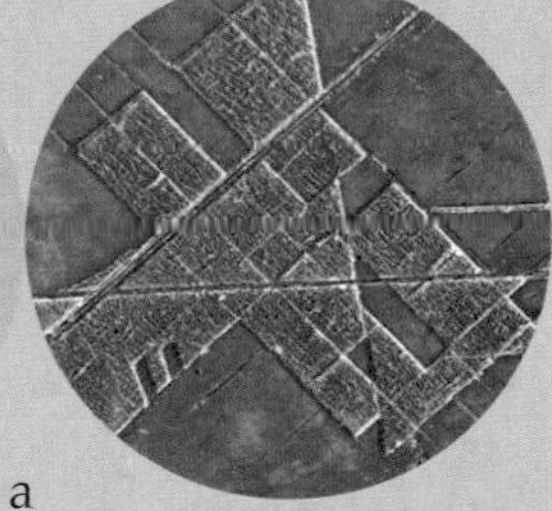

If the fireplace is the perfect complement to a home, then concrete is the perfect material for hearths and fireplace surrounds. It's fireproof, and as we've seen throughout this book, eminently expressive. Its mass holds the emotion, which it will radiate for hours after the fire has gone out, just as a classic Rumford fireplace so beautifully holds and radiates heat.

It's crucial that the design of a fireplace take advantage of the special place it occupies in a home. Most of our fireplaces are unique to their settings and sit comfortably within the space to bring added aesthetic value to the rooms they warm.

« **A FIREPLACE BUILT BY BERNARD MAYBECK** in 1910 for the First Church of Christ, Scientist, Berkeley, uses wood and detailing around it that give a classical feel to an otherwise contemporary composition.

THIS FIREPLACE IS A ROOM within a room in the Ahwahnee Hotel. Part of the magic of this fireplace (besides the giant scale) is that its decorative breaks are tooled in concrete to look like stonework.

In this chapter we'll look at some classic concrete fireplaces that exemplify good design. We'll explore basic concepts of proportion and function; we'll examine how modern zero-clearance metal fireplaces can be made more attractive; and we'll show you how details work to make a piece successful.

Placement and Proportion

When speaking of placement and proportion of the fireplace, we're speaking of relativity. For example, a fireplace so large you can stand inside the firebox may seem ludicrous because it soars beyond the human scale. But in a place like the Ahwahnee Hotel in Yosemite, Calif. (left), just such a concrete fireplace exists. If it were smaller, considering the grand surroundings, it would seem lost and out of place. But with its fine detailing (including a cozy bench to warm your toes); its careful composition of familiar, interconnecting shapes; and its aesthetic blend of textures, the fireplace works.

Bernard Maybeck's fireplaces could be equally majestic, but his designs seem invisibly connected to today's contemporary aesthetic. In looking at his renowned First Church of Christ, Scientist, in Berkeley, Calif., built in 1923, we see how comfortable he was in juxtaposing classical expression and modern style. If you stripped the room of its wood and other detailing, and looked at just the concrete fireplace, the concrete floor, and the metal windows, you might think the place was designed within the past 40 years. And the yet the

BY THE WAY

Some terminology: The "fireplace" refers, as we've used the term, to the entire structure of firebox, surround, and hearth. The "firebox" is the box that holds the fire; the "surround" is the vertical piece around the firebox; the "hearth" is the flat piece in front of the firebox; the "mantel" is the flat top of the surround.

» THIS MAYBECK DESIGN for a residence in Forest Hills, Calif., is stripped of all classical references, save the irons, making it appropriate for a setting that is simple, barn-like, almost Shaker in style.

IN THIS REMODELED HOUSE in Mill Valley, Calif., the fireplace was originally located in the center of the wall. Moved to the corner to accommodate an entertainment center, it occupies the space more comfortably.

fireplace form itself is classical. One sees the echo of Gothic pillars here, and the returns at the base are a feature from a traditional large fireplace. A cook might have placed a pot on one of the returns to warm before dinner. On the other hand, for a home in Forest Hills, Calif. (see p. 165), Maybeck's cast concrete fireplace form seems to strictly follow function. It would blend in with any modern setting.

THE DESIGN OF THIS 20-ft.-long x 5-ft.-high x 4-ft.-deep piece is very stripped-down, but the texture left by rough-sawn boards and oriented strand board gives it warmth and keeps it from looking completely featureless—it is completely contemporary.

THE MASSIVE MANTELPIECE and surround were formed around a metal insert that's a basic 4-ft. x 3-ft. box in this Danville, Calif., contemporary-style home.

CRAFTING GOOD DESIGN

In our work we often have the opportunity to fix hearths and surrounds that simply don't work with the rest of their interiors. In a project in Tiburon, Calif. (see p. 167), and one in Danville, Calif. (above), for example, huge pieces were needed to anchor huge rooms. Metal fireboxes, ones that were woefully undersized, had been specified and installed by the time we came on board to design out and finish the rooms. We used the hearth and surround to provide the mass and proportionate scale necessary to balance out each space. Each hearth is substantial and spans the room. With each surround, the opening flares to the firebox to expand sight lines to the fire.

At the other end of the spectrum, we have a fireplace in Mill Valley, Calif., tucked into a corner beside a large-screen television. In past eras, of course, for many the fireplace was the home entertainment center. The challenge here was to create a comfortable room without both the fireplace and

« **GLASS EMBEDDED** in the concrete surround sparkles when the fire is lit.

^ **THIS NATURAL, GRAY-HUED HEARTH** stretches nearly the length of the wall. The raised and angled edges of the hearth complement the fireplace surround as a sculptural form.

A FIREPLACE WITH A CORRUGATED FIBERGLASS FINISH

Normally used as a roofing material, corrugated fiberglass panels offer interesting possibilities for casting concrete walls. To create the effect, fiberglass panels were fastened to the front wall of the plywood form. To recess the fireplace, form walls were tapered around the opening. Shards of mirrored glass, attached to the tapered walls, were cast into the concrete to create shimmering highlights.

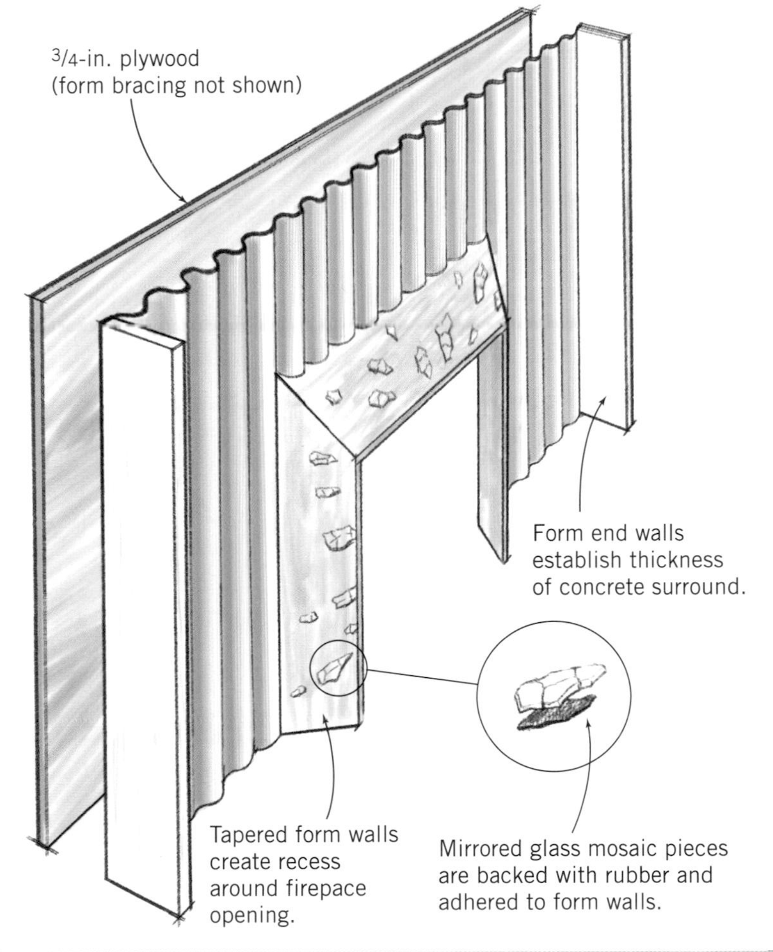

« **MIRRORED PIECES** were affixed with silicone caulk to neoprene rubber inserts. When the form was pulled away, the rubber came with it, leaving the fragments in the concrete.

the entertainment unit competing for attention. We established a hierarchy, with the entertainment unit prominently centered, though discreetly housed in a quiet cabinet, and the fireplace to the side, no longer the central player, but giving warmth and establishing the atmosphere.

The McMillan Project. With the advent of zero-clearance metal fireplace inserts, there's more reason than ever to use concrete for hearths and surrounds. A concrete surround and hearth can ballast these light-weight contraptions, keeping the noble fireplace from becoming an appliance. All too often these days, inserts are simply stuck into a room with no great thought as to whether its size and position make sense. But as the necessity for a fireplace wanes, proportion, scale, and good design remain.

We're often in the position to "make better" the look created by a metal insert—it's something we do gladly. The fireplace surround and hearth, after the lit fire itself, are the most important visual components. The McMillan residence in Danville, Calif. (see p. 168), is a good example of how making over the surround can result in dramatic change. The polished, blue-gray surround for this metal insert was poured on-site using a form lined with corrugated fiberglass. Shards of mirrored glass embedded in the flared opening reflect flames and add unexpected levity to the form. The raised hearth is a long slab that floats, yet connects the fireplace and its grand scale to the room.

YOU CAN BUY a generic, commercial pre-cast fireplace surround and hearth like this one or have one custom-made for your home.

The Pre-cast Surround. Commercial pre-cast concrete fireplace surrounds are widely available. They come packaged in manageable sections that are attached, seamed, and grouted together on-site. The designs are meant to appeal to a wide audience, so many resemble classic stone and marble fireplaces.

Custom crafting a fireplace surround, on the other hand, requires a clear understanding of the individual firebox. For example, some models have small motors inside that blow heated air through ducts for more efficiency. Provisions have to be made for vents in the concrete, and those details must be designed with some forethought to look decent.

A Fireplace in Tiburon. Since our emphasis has been to "beef up" the proportion and size of the typical fireplace, we often cast hearths and surrounds on-site when feasible. Usually the surround is formed and cast first, while the hearth is poured

⩓ **A 16-FT. BLACK,** curved, cast-in-place partition wall in the entry of the Tiburon house segregates the living room fireplace from the front door.

THE CURVED 16-FT. WALL serves three functions: It defines the living room area, diverts traffic from the front door, and provides a niche for firewood and a cabinet.

after the form has been stripped from the surround. This way the braces necessary to secure the forms for the surround can be fastened to the subfloor. The hearth, being flat, won't require as much bracing.

The illustration on the facing page shows how we form the surround around a zero-clearance metal-insert firebox. This insert is positioned in front of the wall, with the chimney exposed. This requires a surround that is massive in appearance, like the surround on the Tiburon job. But often, as with the McMillan fireplace, the insert and chimney are set outside the house, and the firebox opening is flush with the interior wall, so the surround can be thinner.

For the Tiburon project, with the metal insert in place, we first built large knockouts and placed them on either side of the firebox. These large knockouts are constructed of ¾-in. plywood and thoroughly braced inside with 2x6s. They're wrapped with building paper and left in place. We wrapped ¼-in. foam around the edges of the knockouts to prevent cracking (as shown in Chapter 7). The knockouts act as "filler," creating

» **THE HOLDER** for a box of matches is integrated into the concrete.

^ **THE FIREWOOD NICHE** acts as a transition from fireplace to wood cabinet.

hollows inside the surround to reduce weight. (While the surround appears solid, its walls are only about 3 in. thick.) To shape the flared opening to the firebox, we constructed a knockout of plywood and 2x bracing. We pushed the knockout flush to the firebox and caulked the seam with silicone caulk. The rest of the form is a conventional wall form of ¾-in. plywood faced with a liner of plastic laminate. Since we didn't use form ties, the entire structure was thoroughly braced externally, and the braces were screwed to the subfloor. The concrete was lightly colored with ½ percent black.

Details or accessories like the holder for matches, or a wood rack, integrated into the concrete,

HEARTH, MANTEL, AND FIREPLACE SURROUND

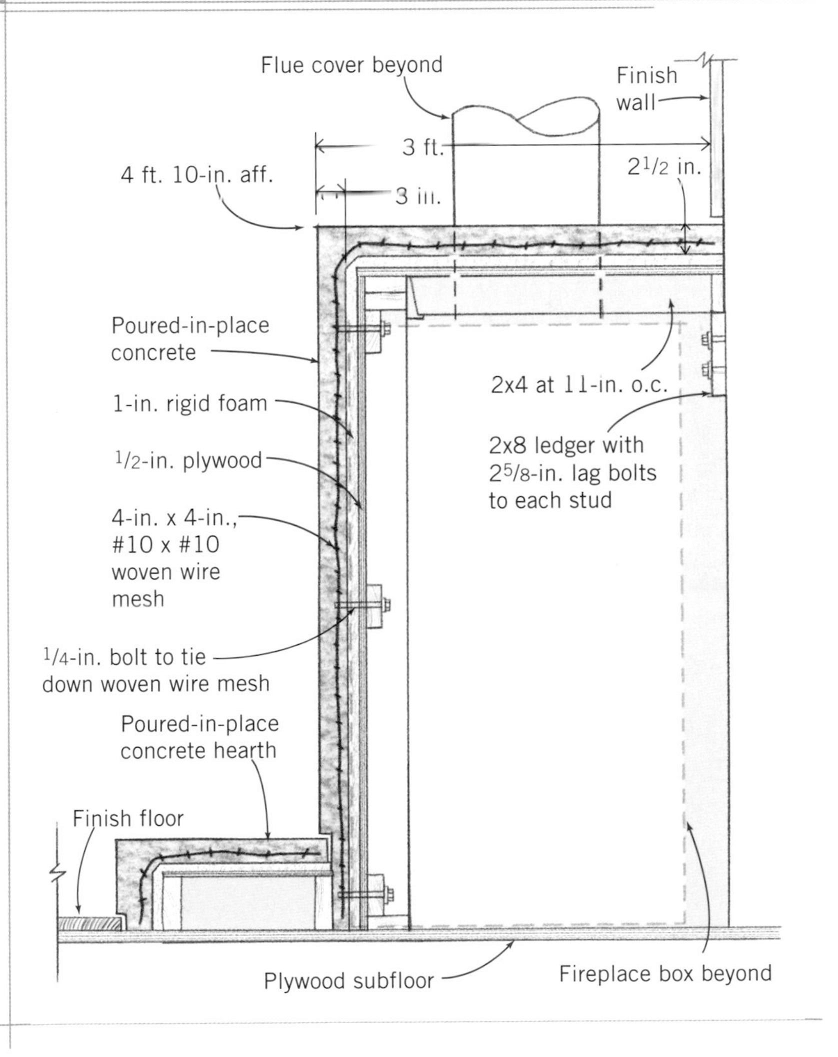

Surrounds from Molds

Dave Condon of Kilnworks in Oakland, Calif., creates custom-fabricated fireplace surrounds by making molds from wood, clay, or plaster originals. His work is classical in style, but his designs are new interpretations of traditional themes. For example, he'll pick up a molding profile and echo the detail somewhere in the fireplace surround. What sets Condon's work apart is that he carves and models in the classic fashion and then uses a rubber urethane mold to reproduce the object in concrete.

1

CONDON STARTED WITH an original terra cotta lattice designed by Maybeck and restored it with new carving.

2

HE THEN POURED a rubber urethane negative mold of the restored sample.

3

THE NEGATIVE RUBBER URETHANE MOLD is framed in wood to the desired module size and sprayed with form release. Beneath is the original terra cotta.

A CONDON-CAST CONCRETE HOOD in this cavernous kitchen resembles an old traditional fireplace cooking hearth.

A NON-SHRINK, non-aggregate gypsum cement grout is poured into the new rubber urethane mold.

A NEW INTERPRETATION of a traditional surround carries an authentic carved stone look using concrete. Classic details are sourced and reconfigured into molds and cast.

provide opportunities for design. The inset for the matches shows off the depth of the concrete by cutting into it. One can get a similar effect by using negative spaces for something like log storage or tools.

Details

If, as the saying goes, the devil is in the details, what better place for him than the fireplace? In the Park City, Utah, project (right), we inserted an Idaho turquoise flat rock through the forms and cast it in for a seat. At the Natoma Street loft in San Francisco, the den has a simple cast-in-place hearth, but the metal corner fireplace insert has been customized with a wood and kindling niche of plate steel.

THE TROTTER FIREPLACE features a stone log holder, a detail that's not difficult to create in concrete. We found a rock at a local landscape yard, cut the bottom flat, and dowelled and anchored it in place.

THE STONE "SEAT" seems magically cantilevered, but the concrete had to be cast around the stone to create the detail.

BY THE WAY

If you choose to cast an entire fireplace in concrete, including the firebox itself, remember that you have to line the box with insulation and firebrick. Regular concrete, unprotected, will crack and disintegrate under continual assaults from fire, creating a potentially life-threatening hazard.

There are concretes designed specifically for high heat, and these are an option. In any case, if you're planning to cast a firebox with any type of concrete, consult with a fireplace expert.

Simplify

This fireplace in a penthouse in Sweden (see p. 182) makes possible the primal experience of the open fire. It's a juxtaposition of the contemporary and the primitive—the design is modern, stripped to the essentials, but the fire itself, set back in the corner against a wall, conjures up the cave.

In the United States, an open fireplace like this one may not pass code restrictions. Rather, fires are often constrained behind doors to save energy, and much of the time all you can see of the fire is a bit of flickering through framed glass. But that shouldn't stop us from looking around for inspiration, from trying to achieve such clean and simple design ourselves—in a mountain cabin, say, anything's possible.

BY DESIGN

SOMETIMES, for reasons beyond our control, the job goes bad. This fireplace surround was poured in place. The driver miscalculated the load, and ran out of concrete before the pour was finished. Since I knew there was no way to accurately match the color, the inevitable cold joint would show no matter what, so I decided to emphasize the joint and turn it into a design element. I stuck in little bits of rock and beach-washed glass, and jammed in some uncolored concrete so the two colored batches were visually separated, then wire-brushed the joint to reveal more aggregates.

» **AT THE NATOMA STREET LOFT,** a concrete hearth was poured in place beneath a corner steel fireplace insert with custom steel wood storage, and plaster above. Only the hearth is concrete, but it shows how a simple slab can create an impact.

Making Lemonade from Lemons

We learned the hard way not to use thin plastic to protect our forms during a pour after a worker on the Natoma Street penthouse in San Francisco used some. The force of concrete pumped up four stories was so great that when it hit, the plastic was sucked into the form along with the concrete. The result was, at first glance, disastrous, but we filled the gaping voids that resulted with a mix of a slightly different color, then ground and polished the hearth to produce a very attractive finished piece, with which everyone, the client included, was quite pleased.

THE SAGGING CONCRETE, gaping holes, trapped plastic, and bits of foam poking through were the grim result.

ALONG THE FRONT, voids were created by the folds in the thin plastic sucked into the form.

WE FILLED THE VOIDS with a mix of slightly different color and smoothed the surface with a damp sponge. (Note the foam knockout over the tile inlays.)

THE HEARTH EXTENDS about 10 ft. Our patch was about a 6-ft. section. After the patch had cured, we ground and polished the entire hearth.

BY BACKFILLING WITH SIMILAR, complementary shades of color slurry, the "natural" void deformations caused by the plastic become a dynamic pattern on the face of the hearth—an accident that could be the basis for another decorative "technique."

THE "OLD" CRACKED PORCELAIN inlay is long and narrow, in keeping with the form of the piece.

» **IN THIS SWEDISH INTERIOR,** the remarkable simplicity of the fireplace complements the serenity of the clean, spare furnishings.

THE WATTS FIREPLACE

Remodeling the fireplace (right) in this Portola Valley, Calif., home took a "less is more" approach. The original—top heavy, with huge inlaid stones and spindly pipe-column supports—looked unstable. It was reconfigured with a balanced composition of materials. We added a steel I-beam, shaved the bulging stones flush, plastered the wall, and covered the hearth with plate steel and concrete to refine the feel of the fireplace without sacrificing its rustic charm.

« **FROM HEAD-ON,** the fire appears to sit in the room. The fireplace's understatement contrasts nicely with the warm textures in the room.

BEFORE ANY WORK WAS DONE, it was evident that the look of the new Watts fireplace would need to be more balanced and refined than the original.

AFTER THE WORK WAS COMPLETED, the visual weight of the fireplace was redistributed. The new hearth hides the stone and anchors the piece to the floor. A ⅜-in.-thick steel plate caps the concrete hearth all around and climbs up to flank the back wall of the firebox.

THE RENOVATION WORK WAS TOUGH—stones couldn't be removed, so they were smoothed down, and a new I-beam was added underneath them.

ARCHITECTURAL ELEMENTS:

9 Water Pieces, Columns & More

We began our book with a focus on work that engages the horizontal and the vertical—floors and walls and countertops—and we moved toward increasingly sculptural and three-dimensional objects such as fireplace surrounds and hearths.

Now we turn to work that takes full advantage of concrete's sculptural qualities and moves from practical objects into yet another dimension—where concrete serves up water and art. We also see how we can bring sculptural concepts to structural pieces, in columns made with fabric forms. And finally, taking our thesis to its logical end, we consider concrete as artwork.

It is often that art takes its inspiration from the elements. In creating water pieces, we look to the source. Water is pure, elemental, and resonates within us all. Unleashed, water can be monstrously destructive. But when water is confined and controlled, its trickling sight and sound hold universally soothing, gentle associations: Water is life, after all.

The Chinese have for thousands of years recognized the benefits—metaphysical and otherwise—of flowing water. In the right place and the right configuration, water will bring on good *qi* (energy, a.k.a. "*chi*"), and subsequently, good

« **WATER GENTLY RUNS** to the end of the channel and down the armature before reaching its destination below, a rice mill from China.

THIS SHALLOW, ROUND BASIN was designed to trap water from the pool so that when it evaporates, mineral deposits remain.

fortune and wealth to the household. These *feng shui* principles have entered popular culture and commerce, and thus we now have recirculating desktop stone-and-water mini-fountains in homes, restaurants, and offices, all faithfully spooning out bad qi and bringing in the good (see the sidebar on p. 189).

But the tabletop fountains are akin to bailing out the Titanic with a teaspoon. I've always been fascinated by large outdoor water features, such as the perpetually flowing, spring-fed fountains that spill over, their waters snaking through the villages in Provençe, France. I decided to bring the same sensibility indoors, combining feng shui and the art of concrete. But I wanted to build something that could move some serious qi around the house.

So I designed several large indoor water pieces for my clients that use concrete to its best advantage. These pieces are produced with the same techniques that we've shown in previous chapters on walls, floors, and countertops. Technically, they are not complex. However, conceptually, we need to consider, for example, how to direct the water so its sounds soothe rather than irritate.

The movement of water through a water piece can be managed in a variety of ways: forming pools, creating streams, allowing the water to percolate or pulse or trickle down. Each approach will form the character of the piece and set the mood in the space. But the most critical factors are that the water must not stagnate and must be replenished periodically. And according to the rule of feng shui, it is always best if the piece can take water in the direction from the outside to the inside rather than vice-versa (it's not a bad idea to consult your local feng shui practitioner to make sure you have the qi going the right way).

Here we'll look at three California projects that illustrate a couple of variations on the water-piece theme. One of these is located at the Téance/Celadon Tea Shop in Albany. Another is in the living room of a home in Los Altos, and the third, a wall in the entry of a home in Palo Alto.

The Tea Shop

Patrons sip their oolong and jasmine teas while contemplating a slow wave of water pulsing down the shallow, slate-lined channel in the middle of the bar-counter at the Téance/Celadon Tea Shop. The wave, perhaps ½-in. high, is generated by a mechanical gate that releases the rising water behind a reservoir every 10 seconds or so. It

THE MECHANICAL WAVE MACHINE creates a pulse-wave with a wheel-cam that lifts a gate in front of a reservoir every ten seconds.

THE MOSAIC-LINED CHANNEL of the water piece at Téance/Celadon was bridged with "boulders" that soften the channel's sharp "shoreline."

travels more than 16 ft. to finally spill onto an armature sluice that conducts the water to an antique milling stone from China. The effect is mesmerizing, and it seems to slow the pace of everyone who sees it.

The technical challenge for us was to craft the sloping 4-in.-wide channel down the middle of the piece without weakening the structure of the 3-in.-thick concrete. So we reinforced the concrete with 1½-in.-thick marine-grade plywood secured under the countertop so it straddles the channel by 8 in. on each side. We then used construction adhesive, which is waterproof, and self-tapping concrete screws (tap-cons) to fasten the plywood to the underside.

Aesthetically, the challenge was what to do with the water once it reached the end of the counter. Would it disappear into a hole in the counter? Would it drop off in a waterfall into a pool? I decided to allow the counter to cantilever off the end of the base, but I held back the drop-off

DESIGNED TO COMPLEMENT AND BALANCE THE EXISTING CONCRETE FIREPLACE, this Los Altos, California, water piece performs as sculpture, guard rail, and room divider separating the upper entry mezzanine with the lower living room. It totals 22 ft. long, with a 14-ft. copper sluice, and was formed and cast in place.

INSPIRED BY NATURE without imitating it, elements of rock and acid-washed concrete are combined with water to create soothing qi.

BY DESIGN

FENG SHUI, literally "wind and water," is the art of "geomancy," or the placement and orientation of buildings in space, or of rooms or objects in those buildings. Feng shui is based on the theory of qi (or chi), which is thought to be a life force that flows through all things. Blockages of qi in the home, due to poor design or inauspicious placement of objects, can cause various disharmonies among the residents, general malaise, illness, and bad luck.

Water is a central element in feng shui. Flowing water enhances the flow of qi, and in particular brings wealth, thus you'll often see a small fountain or aquarium perched by the cash register in a Chinese restaurant. Flowing water into an entryway at home is thought to be especially conducive to harmony and wealth.

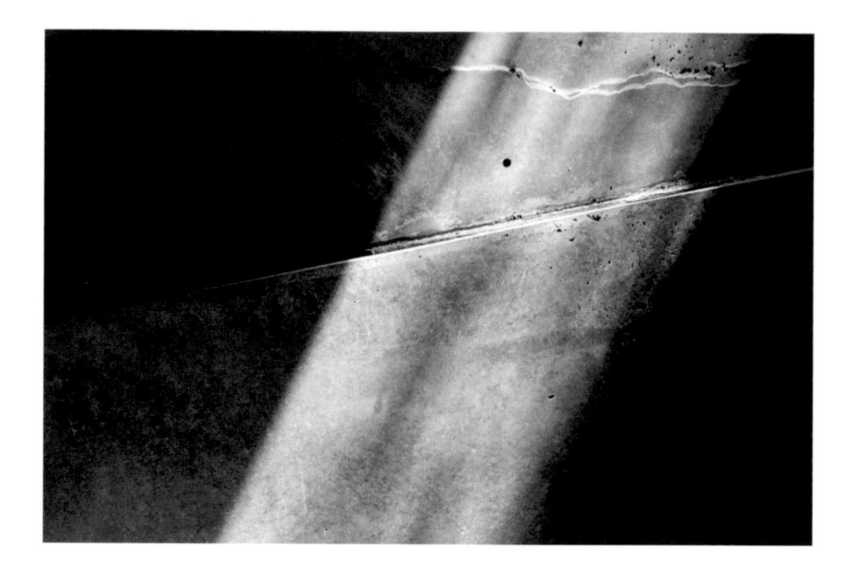

LIGHT CAPTURES the topography of this wall surface.

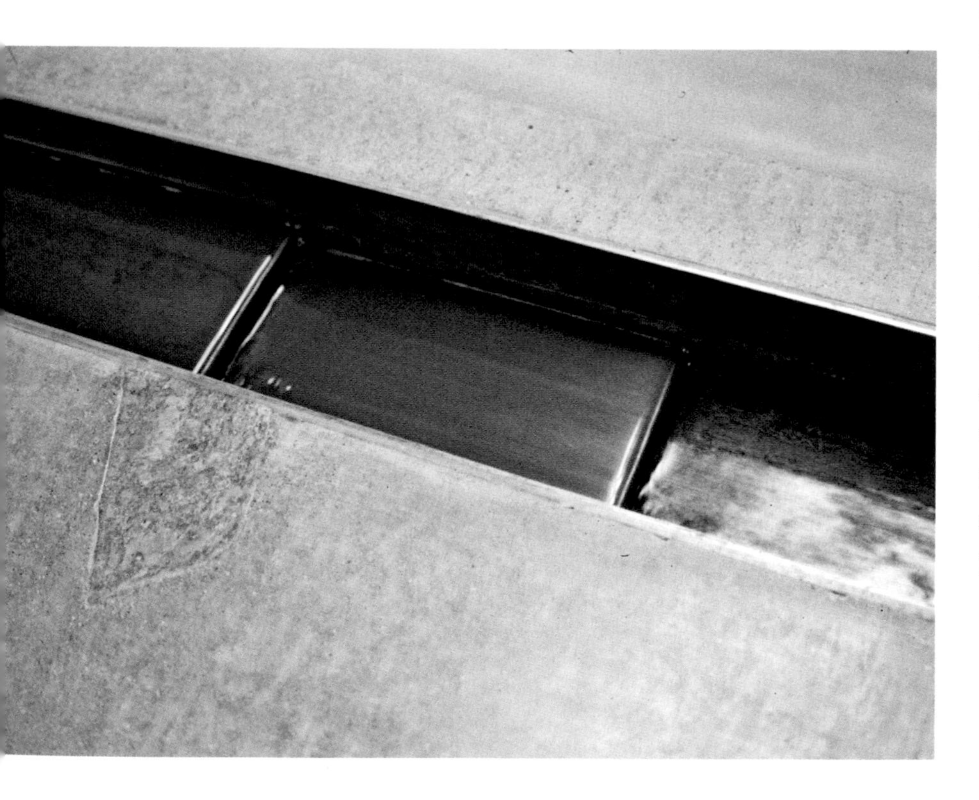

THE COPPER STEPS slow the water's progress in preparation for the descent to the reservoir.

point. This move, in effect, "sculpted" the end of the counter and gave it more dimension.

I wanted the end to be delicate. After considering a metal sluice I came across translucent sheets of beeswax at a candle shop. I gently torched lengths of the sheets so they became warm and pliable, almost melting them onto stainless wire screen to form a sluice. The final touch was to attach my signature, an ammonite fossil, to the end so that the water could coalesce around it and spill in a more controlled fashion into the stone vessel below.

THE LOS ALTOS WATER PIECE

Like the Celadon piece, the water sculpture in the Los Altos home has a long channel with a wave pulse, but it is on the top of a 22-ft. wall. An interior wall of such length needs proper engineering; in this case, fortunately, there was a grade beam under the floor, so the wall was adequately supported.

The first 14 ft. of the channel in the piece are lined with 3⁄16-in. copper plate on the bottom and sides, which was soldered to be water tight. Underneath the copper, the concrete wall was split down the middle with a control joint. This was to account for any differential movement, since the wall sits on two levels. We were also worried that a wall 22 ft. long would likely bend at a weak spot, so a vertical control joint was inserted at a 7-degree angle off the floor.

⌃ **WE TOOLED OUT A SLOT** in the rock to create a little hollow for the electronically-controlled water-wave gate.

BY THE WAY

Water *is* life... and sometimes it's life you don't need, such as algae or mildew, which can cause a rotting smell. If the water in a water piece isn't replenished periodically and the piece isn't cleaned regularly, buildup and decay can leave you with a fetid water piece. The design should accommodate regular maintenance and easy cleaning. Most outdoor pond supply houses have the filters and chemicals to handle algae and bacteria growth. We always design our water pieces with an overflow system, so that (inevitable) mechanical failure doesn't result in a pool of water on a client's floor.

⌃ **THE LONG CHANNEL** sits atop a 22-ft.-long wall.

SECTION OF THE LOS ALTOS WATER WALL

Behind the scenes of a simple concept can be a complex installation. This cross section illustrates how the idea of an unseen source of water in the wall triggered structural changes such as the glue-lam beam and stainless-steel square tube frame required to hold the plaster wall intact with the water tray and riffle board inside the wall.

I always try to build in redundancy with water around. Ostensibly the concrete pit could contain water, but we installed a seamless, welded, stainless-steel tub liner as the holding tank.

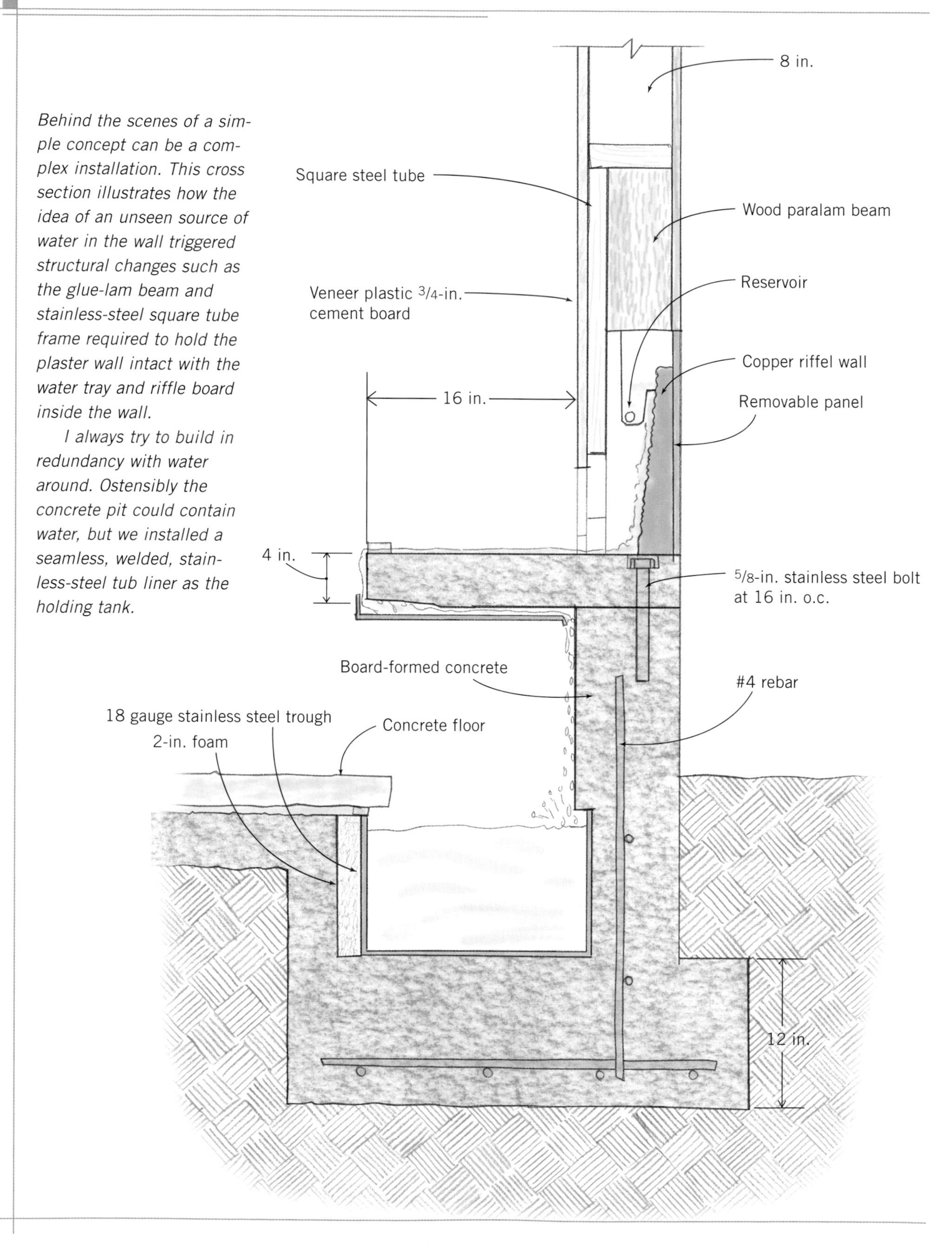

^ **THE WATER PIECE** in the entryway to this Palo Alto home is designed with stillness in mind, an invitation for the hand to skim a finger over the shallow water.

THE PALO ALTO ENTRYWAY

This was a complex installation. The client asked for a water sculpture in the entry, but there was not much room to work in a long piece. However, there was a wall separating the entry from an informal office off the kitchen, and I decided to integrate the water piece into the wall. I played around with possibilities and decided that the "mystery" in this piece was the echo of trickling water without an obvious source. Ultimately, the water runs out of a reservoir hidden inside the wall, down a ribbed metal panel into a shallow pool. Some of the water flows from the pool into a round basin, where it evaporates, leaving mineral deposits. But most of the water spills out over the edge to a brass drip pan, which directs it under the water platform to trickle down the wall to the pool below (see p. 194). From there, it circulates back up to the tank.

To make the piece, we began with the entry floor. I wanted a concrete wall to rise out of a pool in the floor to support the cantilevered water

THE RIBBED CONCRETE in the spillway of this piece was formed in the mold and then wire-brushed after the forms were removed to expose some of the aggregates and provide a visual bridge from the glassine surface of the water to the textured wall below it.

THE PIECE SEPARATES the kitchen and a small office from the public view of the entryway. Water quietly riffles down the wall to the pond in the floor below.

« **THE PIECE'S DELICATE TECTONIC COMPOSITION** is revealed from above: the replica of Mont Blanc, protruding like an atoll in a rectangular sea, is balanced by the shallow circular pool.

feature, so we cast a rectangular concrete pool with the 3-ft.-tall wall rising out of it. We lined the pool with a leak-proof, stainless-steel liner like a large sink (see the drawing on p. 192).

The upper portion of the piece, essentially a countertop, was poured in a mold in our shop. The concrete was colored with black and ultramarine blue pigments, and we embedded a number of items in the concrete, including a small replica of Mont Blanc, which rises up out of the pool like an island.

« **THE RELIEF** hints at an aerial view of an alien civilization.

THE BEAUTIFUL CAPITAL on a column in the Christian Science church designed by Bernard Maybeck shows how he was able to execute intricate design on a large scale using concrete.

The upper wall was constructed of plaster over a wood frame. The reservoir and pumping mechanism are inside the wall; access is through a closet off the office. A pump circulates water through the system, and the reservoir has a ball-cock mechanism that regulates the water level, automatically replenishing the water supply as it drops due to evaporation.

Columns

Before the advent of building materials that permit the design and construction of vast unsupported spans, like steel-reinforced, pre-stressed concrete and steel I-beams, columns had a more important place in design. They were much more common in all settings and usually designed to be beautiful as well as functional.

These days, most columns are purely functional and unadorned, simple round posts formed in the ubiquitous cardboard Sonotube®, or square posts formed in steel or wood forms. Now and then you might see a modest design flourish on a row of columns supporting a freeway overpass, but we hardly notice such things while driving.

Adorned or not, a column is a bit more complicated than it might seem. Its job is to gather forces from a load above and transfer them down to the ground or a foundation. Thus at the top of a column, there might be a capital that flares out to gather the weight. The column itself takes the weight and transfers it to the ground or a foundation. The base distributes the load. A column that will support a load needs to be engineered, but there's no reason why we can't learn from the past and fabricate columns, capitals, and bases that are sculptural and beautiful as well as functional. And concrete is the perfect material for this.

THE SEBASTOPOL FARMHOUSE

On the Sebastopol project, a home styled in the traditions of American and Japanese farmhouses, the living, kitchen, and dining areas comprise a single open space. We used a double row of columns to create a "hallway" that divides the space into two rooms. These columns support a long span of timber framing across the living and dining rooms. The scarf joint connection details between the wood and concrete column bases echo the same joints in the beams overhead.

BERNARD MAYBECK'S CONCRETE SUPPORT for a wooden trellis is an elegant example of how fusing materials makes for aesthetically pleasing and solid construction.

Each column contains a 3-in.-square metal tube as the core. By using steel, we were able to keep the columns small, about 5 in. on a side; an all-concrete column would have had to be more massive to meet the structural engineering requirements.

The concrete post bases are decorative and functional: With small children in the house, the soft fir would have been vulnerable.

THE MAIN SPACE in the Sebastopol, Calif., farmhouse, the author's first commission, is visually divided into living room and dining room by columns and wooden flooring.

THE WOOD-CLAD COLUMNS, with purple integrally colored concrete bases, are made up of four beveled sections of ¾-in. plywood mitered at the corners, then biscuit-jointed and glued.

A FEW VARIATIONS ON THE COLUMN THEME

Dave Condon's columns are at the other end of the design spectrum. Most of his pieces are restorations or new pieces designed to match an existing traditional motif. Fabrication is a simple process; the trick is the artistic talent needed to execute the design.

Condon usually begins with photos or drawings of an original design to be matched. He'll then carve a model in plastic modeling clay (or sometimes wood or wax). Next, he puts the model in a box and pours a two-part rubber mold compound around it, or he'll paint several layers of the compound onto the model. Typically, Condon will wrap the mold in a plaster cast.

Mold and cast are then cut and separated from the model, reassembled, and readied for the pour. The concrete is poured into the mold and allowed to cure, then the mold is removed (it can be reused a number of times).

When the concrete comes out of the mold, it is monochromatic with a slick, plastic look, so Condon will sandblast or acid-etch the surface to give it some texture. This brings out some of the color of the aggregates and fines, giving the piece more subtlety and interest. Sometimes Condon will add more color by sponge-brushing the piece with a slurry of cement, pigment, and fine sand.

« **DAVE CONDON USES MODERN MOLD MATERIALS** and techniques to cast concrete pieces that look carved. Most of his work is in carved restorations, or pieces such as the column capitals, designed to match an existing motif. Note the texture, produced by sand-blasting.

1 **CONDON STARTS BY CARVING THE MODEL** in plastic modeling clay, though he may use wood, wax, or whatever is appropriate for the piece. The completed piece was painted with several layers of a two-part rubber mold compound (see Resources, p. 208).

2 **ONCE THE PLASTER HAS FIRMED UP,** he'll cut the mold away from the model, put the two pieces back together, and fill with concrete. If any seam should show, it can be polished out.

3 **BECAUSE THE MOLD IS SOFT AND FLEXIBLE,** it needs a stiff backing so it can hold its shape after the model is removed. In this case, Condon has wrapped the mold in cloth and plaster, like a cast.

HUGGING THE CONTOUR of the hill, the house overlooks the distant Virgin Islands. Made of solid concrete, it is engineered to withstand hurricanes.

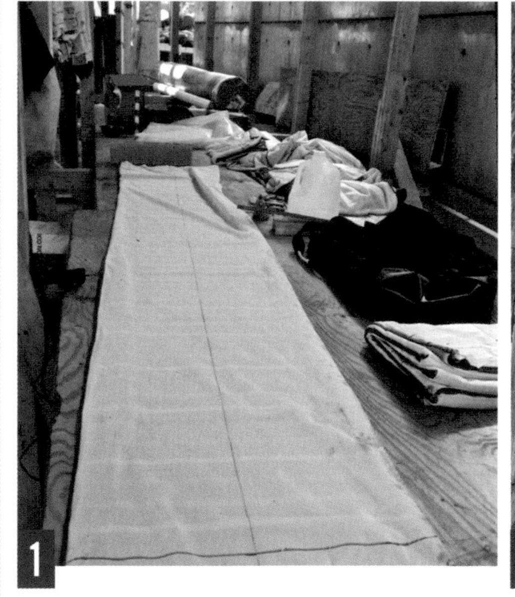

THE SPANDEX® LINER fits like a giant sock over #6 rebar from the floor to the plywood roof deck.

CONCRETE WILL BE POURED INTO THE FABRIC FORMS through holes in the plywood roof deck. Once the columns have been poured and fully cured, more concrete will be poured over the entire deck to create the roof.

THE SPANDEX AND GEOTEX® are pulled and sandwiched in between the collars to seal off the bottom.

THE CULEBRA ISLAND PROJECT

We introduced the concrete walls of this project in Chapter 6, but that isn't the only concrete story to be told. In fact, the entire house was a continuous site-cast pour. It was structurally engineered by Robert Lawson to withstand sustained winds to 250 miles per hour. Because of the complex engineering and design (see model on p. 202), there were few crews on the island with the experience to carry it off. I had to find the simplest systems to build sophistication into the home with locals and local limitations.

To pour the 160-ft. curved wall we avoided form failure by devising the fool-proof, slip-cast system. But pouring the columns hinged on finding the right person with the right *invention*: I desperately wanted *not* to use the standard, soul-less, cardboard Sonotube. Yet, other expen-

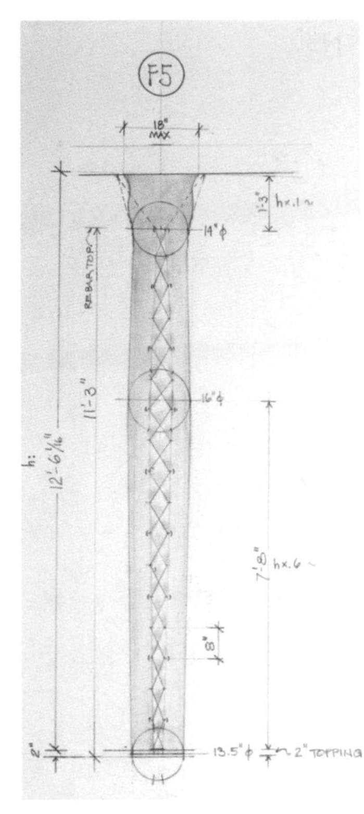

WE POURED THIRTEEN COLUMNS, each with its own nuances of shape and form fashioned by adjusting the laces along the length of the Geotex form. The girdle was removed and the Spandex stripped after a five-day cure. A month later, the columns took on the full load of the 8-in. concrete roof deck.

4

INSIDE, THE SPANDEX is stretched taut. Outside, the Geotex is laced with steel fasteners hooked around rebar into sleeves of Geotex.

5

THE GEOTEX "GIRDLE" is fastened and ready. At the top of the column the Spandex will bulge and form a natural "capital." In the background, note the plywood collar bolted to the floor.

CHAPTER THREE

LARGE PLYWOOD KNOCKOUTS were inserted in the forms to create an elliptical window in the finished wall and a circle in the roof.

WOOD SCAFFOLDS temporarily held up a wood deck that the roof columns were poured from. The roof was poured after the columns cured.

A MODEL OF THE CULEBRA ISLAND, Puerto Rico, house shows how the slip-cast curved walls embrace the home and nestle it into the contours of the hillside.

sive options meant shaping in wood or steel, thirteen different sized forms for the thirteen columns at thirteen points on a sloping and pitched roof.

Then, I recalled having seen an artist's work in fabric-formed concrete a few years earlier. Months later I met Mark West on the island. He had with him the thirteen column forms…in duffel bags! For the next three weeks, in work more akin to pitching tents than building forms, we set and poured the columns.

The Centre for Architectural Structures

Architect Mark West, director of the Centre for Architectural Structures and Technology at the University of Manitoba in Canada, has invented a technology for molding concrete with fabric forms using a combination of elastic Spandex and Geotex. The beauty of the technique is that you can affect the profile of the wall or tailor the shape of a column simply by manipulating the Geotex, which contains the concrete-filled inner Spandex lining. As a result, you've frozen the forces at work in the concrete.

ART, FORM, AND ENGINEERING are synonymous in the works by Mark West.

THE CENTRE'S FACILITY is primarily focused on the exploration of imaginative engineering with concrete.

EACH RAISED ELEMENT on this wall section was created by patterns of tension in the fabric form.

» **THE STRIATIONS** of gray, green, and brick-red in Louis Meyer's trellis column were created by pouring separate batches of integrally-colored concrete into the mold.

^ **LOUIS MEYER'S TREE-LIMB SUPPORT** contains a number of found objects, including plumbing parts and bits of broken brick.

Art Pieces

I have always felt that artists are the research scientists of the design world, discovering through their art new and subtle ways of looking. As a designer and builder, I try to apply the concepts from their research to the everyday environment in a fusion of art, design, and craft. This fusion is what can make work live and, most important, make it fun to produce. What follows are examples from some colleagues (including my brother Carl Cheng) whose work may inspire you as it inspires me to explore the best of creativity and craftsmanship in concrete.

LOUIS MEYER

Berkeley artist Louis Meyer likes to sculpt using found objects and concrete, which, he notes, is a wonderfully cheap material for the thrifty artist. Meyer has produced a number of concrete pieces for an outdoor cafe, including tables and benches, and the two whimsical "columns" shown here, one of which supports a trellis, the other a tree limb. While the concrete was still quite green, he removed the forms, then hit the concrete with a hard spray of water. Finally, he carved the concrete with a variety of steel tools to give it a loose, spontaneous look.

DENNIS LEFBOM

Dennis Lefbom of Function Formworks creates small pieces—candelabras, lamp stands, chess pieces, and planters—with simple materials like rigid foam. His sculptural pieces are typically unpigmented gray concrete poured into forms. Because he captures so much detail in his work, his mix includes a heavy ratio of cement to sand and no aggregates.

THIS CHESS SET by Dennis Lefbom illustrates his facility with small forms.

COMPLETED IN 2000, *Homage to Huntington Beach*, by Scott Donahue, is a low, carved art wall designed for a busy corner and depicts four historical periods of life in the area.

SCOTT DONAHUE

Bay Area public artist Scott Donahue is known for his oversized sculptures. He often chooses subjects that celebrate local history. In classic sculptor fashion, he carves full size clay models then makes sectional molds of plaster and rubber, pours concrete into the molds and, once the pieces are cured, assembles and contours them into the finished sculpture.

FREEWAYS, HOUSES, CARS, AND FACTORIES were made in wood, vacuum-molded, and recast into concrete panels.

CARL CHENG

I have found inspiration in my brother Carl's work—not only in the art concepts that he conceives but in his innovative processes for carrying them out. The entryway wall at the Sebastopol home (see Chapter 5), which has eroded away to reveal a trophy buried in it during the pour, for example, was inspired by a small erosion piece Carl created years before. And the rubber-mold process he used to produce his prize-winning *Walk on L.A.* showed me how easily I could mold concrete in my own work.

Related to the theme of erosion, *Walk on L.A.* is a tool for making art in sand, art that is meant to be tromped on, wiggled on, and sprawled on by the public. Once a month, this roller, 10 ft. in diameter, is pulled along by city tractors at the same time they drag screens to clean the beaches.

THE RELIEF on the concrete roller was cast from rubber molds. The concept for *Walk on L.A.* won the Public Art competition in Santa Monica in 1988.

RECYCLED CONCRETE

Concrete can make ephemeral art and it can make enduring art. Recycled concrete is a new kind of stone. For the walk leading to the Hogan-Mayo house, concrete slabs from a dump were shaped and dry-fitted into a rhythmic pathway border according to a plan by landscape architect Steve Adams. His resurrected concrete is re-born in the hands of a master mason, and as a symbol of good design and innovation, is the perfect place to close the story of concrete at home.

Resources

DESIGN SERVICES, CONCRETE FABRICATORS, AND SEMINARS

CHENG DESIGN
Principal, Fu-Tung Cheng
2808 San Pablo Ave.
Berkeley, CA 94702
(510) 849-3272
www.chengdesign.com

CHENG CONCRETE EXCHANGE
www.concreteexchange.com
An on-line resource store for professionals, homeowners, and do-it-yourselfers which includes a national directory of fabricators, artists, and mold-making supplies.

CHENG CONCRETE EXCHANGE ACADEMY
1142 6th St.
Berkeley, CA 94710
(510) 849-3272
www.concreteexchange.com
Offers basic and advanced courses, and 5-day seminars in concrete countertop making for professionals and do-it-yourselfers. Also available: wall-forming, water pieces, and fireplaces.

INSTITUTE FOR AMERICAN CRAFTSMANSHIP
3410 West 11th Ave.
Eugene, OR 94702
(877) 935-8906
www.tradeskills.org
Basic and advanced technical training in decorative concrete, masonry, painting, and decorating.

CONTRIBUTING ARTISTS

STEVE ADAMS, LANDSCAPE ARCHITECT
Adams Design Associates
143 S Cedros Ave., # V105
Solana Beach, CA 92075
Phone: (858) 350-8166

WENDELL BURNETTE
www.wendellburnettarchitects.com

MICHAEL MAISTER LANDSCAPE
www.maisterlandscape.com
Hogan-Mayo Residence

JOE WEISS MASONRY
(760) 436-9121
Hogan-Mayo Residence

CARL CHENG, PUBLIC ARTIST
John Doe Co.
1518 17th St.
Santa Monica, CA 90404
Johndoe.co@aol.com

DAVE CONDON
Kiln Works
P.O. Box 2573
Castro Valley, CA 94546
Phone: (925) 606-6560
Fax: (925) 606-5478

JIM LUNDY
All Surface Cleaning Systems, Inc.
San Francisco, CA
(415) 716-0690

SCOTT DONAHUE
1420 45th Street, # 49
Emeryville, CA 94608
Email: scott@sdonahue.com
www.sdonahue.com

DALE BLAYONE
Concrete Cuisine
44 West 3rd St.
Hamilton, Ontario
Canada L9C 3J9
905-385-8900
innova@mountaincable.net

MICHAEL KARMODY
Stone Soup Concrete
221 Pine St.
Florence, MA 01062
(800) 819-3456
www.stonesoupconcrete.com

JAMES McGUIRE AND ANDREW SIMON
Lokahi Stone
518 Kamani St.
Honolulu, HI 96813
(808) 596-8833
www.lokahistone.com

COUNTERCAST DESIGNS, INC.
6875 King George Highway
Surrey, BC Canada
(888) 787-CAST
www.countercast.com

LOUIS MEYER
Turn of the Century Furniture
Berkeley, CA
(510) 849-0905

DENNIS LEFTBOM
Functional Formworks
San Francisco, CA
(415) 922-3237

THE CONCRETISTS,
a flat-work artisan collective,
(www.theconcretist.com) with:

MIKE MILLER
The Cottage at 395 West "J" St.
Benicia, CA 94510
(707) 280-7195

KELLY BURNHAM
Clayton, CA
(925) 324-0714

DANA BOYER
Apache Junction, AZ
(602) 369-1792

KAREN TIERNEY
Tierney@theconcretist.com

MOLD-MAKING SUPPLIES

POLYTEK DEVELOPMENT CORP.
55 Hilton St.
Easton, PA 18042
Phone: (800) 858-5990
Fax: (610) 559-8626
www.polytek.com

FLOORS: GENERAL SUPPLIES

ARDEX ENGINEERED CEMENTS, INC.
400 Ardex Park Dr.
Aliquippa, PA 15001
Phone: (724) 203-5000
Fax: (724) 203-5001
info@ardex.com

SURFACE GEL TEK, LLC
310 S. Nina Dr., #10
Mesa, AZ 85210
Phone: (888) 872-7759, (480) 970-4580
Fax: (480) 421-6322
info@surfacegeltek.com
www.surfacegeltek.com

3M (FOR BUTTERCUT SAND-BLASTING STENCIL MATERIAL)
(888) 364-3577
www.3m.com/US/

FLOORS: COLOR HARDENERS, ACID STAINS, STAMPING EQUIPMENT

Most of the companies listed below have a line of proprietary products that work as a system for coloring and stamping floors for professionals:

BUTTERFIELD COLOR
625 W. Illinois Ave.
Aurora, IL 60506
Phone: (800) 282-3388
Fax: (630) 906-1982
email@butterfieldcolor.com
www.butterfieldcolor.com

SOLOMON COLORS
Phone: (800) 624-0261, (217) 522-3112
Fax: (800) 624-3147, (217) 522-3145
sgs@solomoncolors.com
www.solomoncolors.com

KEMIKO CONCRETE PRODUCTS
P.O. Box 1109
Leonard, TX 75452
Phone: (903) 587-3708
Fax: (903) 587-9490
sales@kemiko.com
www.kemiko.com

L. M. SCOFIELD COMPANY
4155 Scofield Rd.
Douglasville, GA 30134
Phone: (770) 920-6000
Fax: (770) 920-6060
www.scofield.com

FLOORS: GRINDING AND POLISHING EQUIPMENT

Companies listed below sell machines, pads, and grinding/polishing tools:

VIC INTERNATIONAL CORPORATION
Phone: (800) 423-1634
Fax: (800) 242-1141
www.vicintl.com

SASE COMPANY, INC.
26423 79th Ave. South
Kent, WA 98032
Phone: (800) 522-2606
Fax: (877) 762-0748
www.sasecompany.com

SAWTEC
6215 Aluma Valley Dr.
Oklahoma City, OK 73121
Phone: (800) 624-7832
Fax: (405) 478-3440
www.surfacepreparation.com/Sawtec.htm.

COUNTERTOP SUPPLIES

CARBON FIBER MESH:
TechFab, LLC
P.O. Box 807
Anderson, SC 29622
2200 South Murray Ave.
Anderson, SC 29624
Phone: (864)-260-3355
Fax: 864-260-3364
www.techfabllc.com

CHENG CONCRETE EXCHANGE
Online countertop specialty store: NeoMix® kits, Pro-Formula concrete countertop mix, sealers, sink and mold knock-outs, cordless vibrators, water-feed grinder/polishers, diamond pads, decorative semi-precious aggregates, and other accessories.
www.concreteexchange.com

Resources

WHITE CAP INDUSTRIES
General supply and tools for concrete work including trowels and hand-tools.
www.whitecapdirect.com

CEMENT MIXERS

IMER USA INC
www.imerusa.com

GILSON MIXERS:
Clearform Tool Co.
4343 Easton Rd.
St. Joseph, MO 64501
(800) 253-3676

WHITEMAN INDUSTRIES
P.O. Box 6254
Carson, CA 90749
(800) 421-1244

HANDHELD EQUIPMENT SUPPLIERS

DUST MUZZLE
837 Cornish Dr.
San Diego, CA 92107
www.dustmuzzle.com

FEIN POWER TOOLS
1030 Alcon St.
Pittsburgh, PA 15220
Phone: (800) 441-9878
Fax: (412) 922-8767

ISKCO VIBRATION EQUIPMENT AND SYSTEMS
www.iskco.com

WYCO TOOL CO.
(stinger vibrators for standard wall forms)
2100 South St.
Racine, WI 53404
(800) 233-9926

METABO CORP.
(hand-held power tools)
1231 Wilson Dr.
P.O.Box 2287
Brandywine, Industrial Park
West Chester, PA 19380
Phone: (610) 436-5900
Fax: (610) 436-9072
info@metabousa.com

MAKITA USA
14930 Northam St.
La Mirada, CA 90638
(714) 522-8088

PIGMENTS

PFIZER
235 E. 42nd St.
New York, NY 10017
(212) 573-1000

DAVIS COLORS
3700 East Olympic Blvd.
Los Angeles, CA 90023
Phone: (800) 356-4848
Fax: (323) 269-1053
www.daviscolors.com

ELEMENTIS PIGMENTS
Corporate Headquarters
2051 Lynch Ave.
East Saint Louis, IL 62204
Tel: (618) 646-2110
Fax: (618) 646-2178
pigments.info@elementis-na.com

TRADE ORGANIZATIONS AND ASSOCIATIONS

AMERICAN CONCRETE INSTITUTE
P.O. Box 9094
Farmington Hills, MI 48333
(248) 848-3700
Provides standards and practices, building code requirements, software, seminars, and certification.

PORTLAND CEMENT ASSOCIATION
5420 Old Orchard Rd.
Skokie, IL 60077
Phone: (847) 966-6200
Fax: (847) 966-8389
www.cement.org
Technical publications, computer programs, and educational materials

THE NATIONAL KITCHEN & BATH ASSOCIATION (NKBA)
687 Willow Grove St.
Hackettstown, NJ 07840
Phone: (800) 843-6522
Fax: (908) 852-1695
www.nkba.org
Organization for manufacturers, distributors, kitchen and bath showroom dealers, kitchen designers, and interior designers.

AMERICAN SOCIETY OF INTERIOR DESIGNERS (ASID)
608 Massachusetts Avenue, N.E.
Washington, DC 20002-6006
Phone: (202) 546-3480
Fax: (202) 546-3240
www.asid.org

CONCRETE RETAIL SPECIALIST

GRANITEROCK
350 Technology Dr.
Watsonville, CA 95076
Phone: (831) 768-2000
Fax: (831) 768-2201
Email: mainoffice@graniterock.com
www.graniterock.com

PRINTED MATTER:

BUILDER'S BOOKSOURCE
1817 4th St.
Berkeley, CA 94710
(510) 845-6874
www.buildersbooksource.com

CONCRETE COUNTERTOPS: Design, Forms, and Finishes for the New Kitchen and Bath
by Fu-Tung Cheng
The Taunton Press, Inc., 2002

CONCRETE DÉCOR: The Journal of Decorative Concrete
Professional Trade Publications, Inc.
P.O. Box 25210
Eugene, OR 97402
(541) 341-3390
www.concretedecor.net

CONCRETE MANUAL
International Conference of Building Officials (November 1998)
5360 Workman Mill Rd.
Whittier, CA 90601
(800) 284-4406

CONCRETE STRUCTURE, PROPERTIES, AND MATERIALS
by P. Kumar Mehta
Prentice-Hall, Inc., 1986

NKBA KITCHEN BASICS: A Trainer Primer for Kitchen Specialists
by Patrick J. Galvin with Ellen Cheever

THE TIMELESS WAY OF BUILDING
by Christopher Alexander
Oxford University Press, 1979

A PATTERN LANGUAGE
by Christopher Alexander
Oxford University Press, 1977

IT'S YOUR KITCHEN: Over 100 Inspirational Kitchens
by Joan Kohn
Bulfinch, October 2003

RESIDENTIAL CONCRETE
by National Association of Home Builders
Portland Cement Association, 1983

MODERN KITCHEN WORKBOOK: A Design Guide for Planning a Modern Kitchen
by Wanda Jankowski
Rockport Publishers, 2001

WABI-SABI: for Artists, Designers, Poets & Philosophers
by Leonard Koren
Stone Bridge Press, 1994

TOTAL DESIGN: Contemplate, Cleanse, Clarify, and Create Your Personal Spaces
by Clodagh Aubry, Heather Ramsdell, Daniel Aubry
Clarkson Potter, 2001

TAO TE CHING: The Way and Its Power
by Arthur Waley

BERNARD MAYBECK: Visionary Architect
by Sally B. Woodbridge, photography by Richard Barnes
Abbeville Press, 1992

Index

Index note: page references in *italics* indicate a photograph; references in **bold** indicate a drawing or illustration.

D

E

F

G

H

I

Index

Credits

All photos by Matthew Millman, except as noted:

p. iii (bottom): Photo by Mike Miller

p. v (top): Photo by Mike Miller; (bottom) Photo by Tim Moloney

p. vii (left, middle): Photo by Richard Barnes; (left, bottom): Photo by Fu-Tung Cheng

p. 1: Photo by Mark Cohen

p. 3 (left): Photo by Richard Barnes

p. 5 (right): Photo by Richard Barnes

p. 8: Photo by Richard Barnes

CHAPTER 1

p. 9: Photo by Mike Miller; p. 13 (left): Photo by Chuck Miller, © The Taunton Press, Inc.; p. 14 (top): Photo by Richard Barnes; p. 19: Photo by Mark Cohen

CHAPTER 2

p. 26, 29 (right), 40, 42 (top), 43 (left), 45 (right, top and bottom): Photos by Robert Ryan; p. 42 (bottom): Photo by Fu-Tung Cheng; p. 47: Photo courtesy Western Floor Specialists

CHAPTER 3

p. 64 (top): Photo courtesy Western Floor Specialists; pp. 72–75: Photos by Mike Miller

CHAPTER 4

p. 80 (top left): Photo courtesy Countercast Designs Inc.; (bottom left): Photo by Mark Foreman; (top and bottom, right): Photos courtesy Lokahi Stone; p. 81: Photos courtesy Concrete Cuisine; p. 91 (bottom): Photos by Fu-Tung Cheng; pp. 92, 94: Photos by Morgan Conger

CHAPTER 5

pp. 99 (detail), 100 (top, left and right), 101 (bottom), 109 (right), 114 (top), 115 (bottom), 117 (bottom), 118 (bottom): Photos by Fu-Tung Cheng; pp. 100 (bottom), p. 101 (top), 104 (left), 105–106, 111: Photos by Richard Barnes; pp. 108–109 (left, top and bottom): Photos by Wendell Burnette; p. 112: Photo by Andre Ramjoue; pp. 114 (bottom), 115 (top): Photos by Tim Moloney

CHAPTER 6

p. 124: Photo by Richard Barnes; p. 126: Photo by Mark Cohen; pp. 131 (top left), 142 (left), 143: Photos by Fu-Tung Cheng

CHAPTER 8

pp. 162, 165, 166: Photos by Richard Barnes; pp. 167, 173 (left and right), 183 (right, top and bottom): Photos by Fu-Tung Cheng; p. 170: Photo courtesy Sierra Concrete, Inc.; pp. 176 (right), 177: Photos by Andre Ramjoue; p. 182 (top and bottom): Photos by Ake Lindman

CHAPTER 9

pp. 185 (detail), 188, 189, 196–197, 203 (detail, top left): Photos by Richard Barnes; pp. 190–191, 200 (top), 202 (top left): Photos by Fu-Tung Cheng; p. 198 (top): Photo by www.davidduncanlivingston.com; p. 200 (top): Photo by Fu-Tung Cheng; (bottom, left to right), 201 (all), 202 (bottom), 203 (all): Photos by Mark West; p. 202 (top left and right): Photos by Fu-Tung Cheng; p. 205 (right): Photo by Dennis Lefbom; (left): Photo by Scott Donahue